EXPLORING THE
SOLAR SYSTEM

EXPLORING THE
SOLAR
SYSTEM

NICHOLAS BOOTH

CAMBRIDGE
UNIVERSITY PRESS

ACKNOWLEDGEMENTS

Any book that attempts to record the remarkable legacy of three decades of planetary
exploration lives or dies by the quality of pictures within it. The help and sheer
professionalism of the "picture providers" within the public information offices of NASA
and ESA has been exemplary. Without them, this book would not have been possible,
and I would like to express my thanks to Mike Gentry and Becky Fryday at NASA Johnson,
Jurrie Van Der Woude at JPL, Ray Villard at the Space Telescope Science Institute and
Simon Vermeer at ESA in Paris.

My thanks also to Bill Hemsley who nursed this book through protracted labour pains and
my agent, Vivienne Schuster, who helped spirit them away. Paddy Tate and Karen Stewart
were inspired in their design work. And thanks also to Al and Marka Hibbs for allowing me
to write parts of this book in Pasadena. To all of you, many, many thanks.

NICHOLAS BOOTH
April 1995

For all JPLers everywhere – especially Al and Marka
Hibbs, Dean Kastel, Lorna Griffiths and Don Bane,
who made my own journey of exploration possible.

Published by the Press Syndicate of the University of Cambridge
The Pitt Building, Trumpington Street, Cambridge CB2 1RP
40 West 20th Street, New York, NY 10011-4211, USA
10 Stamford Road, Oakleigh, Melbourne 3166, Australia

Copyright © Reed International Books Ltd 1995

First published in Great Britain in 1995 by George Philip Limited

First published by Cambridge University Press 1996

Reprinted 1998

Library of Congress cataloguing in publication data available

This edition only for sale in the United States of America
and Canada

ISBN 0 521 58005 6 hardback

Printed in China

CONTENTS

INTRODUCTION

Welcome to the myriad worlds of the Solar System, revealed in stunning detail by robotic envoys from Earth. Only a few decades ago these planets and moons were mysterious, distant worlds, which shimmered tantalizingly through telescope eyepieces. Today they are landscapes as vivid and real as those of our own planet. Once they provided fertile grounds for writers' flights of fancy; now we know them to possess a staggering range of features surpassing anything imagined in science fiction. Planetary exploration has vastly expanded our hori-

Earthrise *Undoubtedly one of the most memorable images of the golden era of planetary exploration is the view of Earth rising above the Moon's horizon. The photograph was taken from Apollo 11 in July 1969. The terrain is in the area of Mare Smythii.*

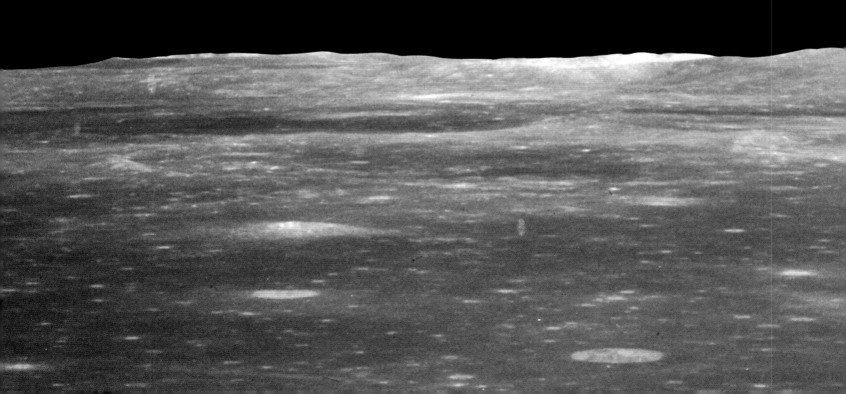

Farewell to Earth

The Moon and Earth (right), taken from the Galileo spacecraft as it looked back from a distance of 6.2 million km on its way to Jupiter in December 1992. The colour-enhanced image was processed by computer.

zons; ours has been the first generation privileged to see the planets in all their glory and diversity. Our scientific and philosophical perspectives have been enormously expanded and, for the first time, we are beginning to have a clearer and more humble idea of our place in the Universe.

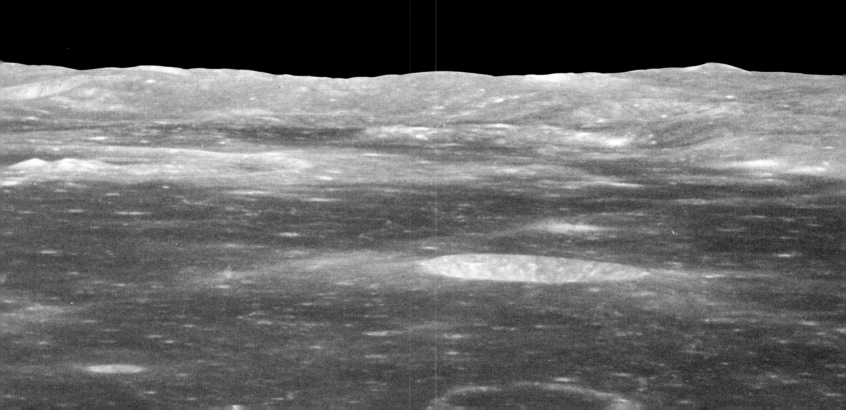

◄ **Mountains on Venus**
Maxwell Montes (just above centre) is the tallest mountain range on Venus, rising 11 km above the average terrain level. The image was generated by computer from data gathered by the Magellan radar-mapping spacecraft. Colours have been added, as well as clouds, mist and atmospheric haze.

Since the early 1960s, handfuls of crewless probes have journeyed to every planet in the Solar System except distant Pluto. The first missions in the 1960s simply flew past the objects of their scrutiny, while later probes have orbited and examined the planets' surfaces in detail for months at a time. A dozen humans have walked on Earth's Moon, and probes have landed on our planetary neighbours Venus and Mars. In the late 1970s and the 1980s the giant worlds of the outer Solar System, with their curious rings and retinues of moons, were observed from close range by the two Voyager spacecraft – perhaps the most successful missions to date. Small wonder that planetary researchers refer to this period as the golden age of planetary exploration.

We have now entered a second stage of our research of the planets, although in the current economic and political circumstances it is hard to imagine the missions now planned acquiring anything remotely resembling a golden aura.

The images from the golden age comprise what is still a priceless electronic archive. With new computer techniques these images have been processed and enhanced, giving us, for example, remarkably realistic three-dimensional views of planetary surfaces, scaled to the correct perspective. The resulting visual bounty has, to a great extent, made this illustrated book possible.

► **Ocean world** The Earth's surface is 70 per cent covered with water, and the oceans seem to play a dominant role in determining the Earth's weather and affecting global climate change, although the processes are still not fully understood. The photograph here was taken during the Space Radar Laboratory mission on NASA's space shuttle.

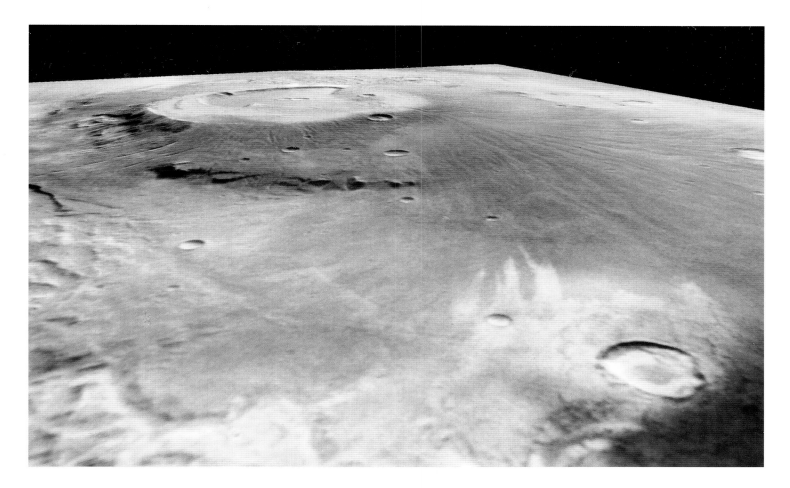

Space Age

The opening up of the Solar System has provided us with more than just pretty pictures. Besides cameras, all planetary spacecraft have been equipped with instruments sensitive to exotic radiations and particles invisible to the naked eye. The Sun, for instance, constantly emits a stream of charged particles which interact with the planets and their magnetic fields. Using techniques originally developed by particle physicists, we have been able to discover a range of curious phenomena resulting from the Sun's influence throughout the Solar System. These observations have given us insights into how planets interact with the Sun.

It has been said that the Space Age came at exactly the right moment for planetary astronomers. By the end of the 1950s, there was little new information to be gained from looking at distant planets with the technology then available at even the best astronomical observing sites. Our atmosphere presents a frustrating barrier to ground-based observation because it stops a good proportion of the full range of electromagnetic radiation (which includes visible, ultraviolet and infra-red light, X-rays and radio waves) from reaching the Earth's surface. If we wanted to learn more about the planets, the only way to do it was to go into space ourselves. The International Geophysical Year

▲ **Mars-scape** Mars, along with Venus, was the focus for the first wave of planetary exploration in the 1960s and 1970s, but latterly missions have been jinxed by a series of heart-breaking failures, which culminated in the loss of the two Russian Phobos probes in 1988–89 and the United States' Mars Observer mission in 1993. Planetary scientists have nevertheless been able to produce startling new views of Mars with realistic perspective, created by using three-dimensional models based on the old Viking-mission data and photographs.

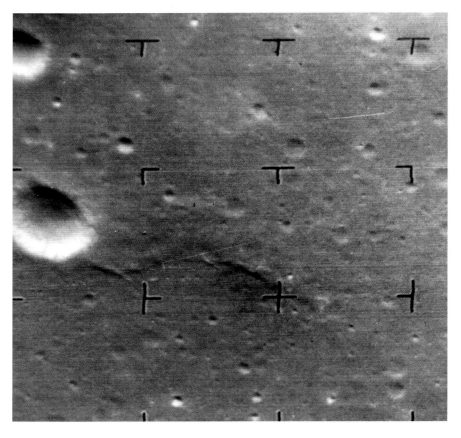

The United States' first satellite, Explorer 1, was equipped with a Geiger counter and discovered belts of trapped radiation around the Earth. For the first time it was possible to get a truly global perspective on our planet. As a result, a fundamental revolution in science has taken place which may be gauged by the diversity of research it has stimulated. Traditionally separate disciplines of scientific endeavour have become closely allied as we have worked to understand the Solar System as a whole.

Geologists ponder how and why the surfaces of other worlds look as they do, scarred by impact craters, some with signs of volcanic activity. Meteorologists watch in awe as storms the size of the Earth swirl in the atmospheres of the outer planets. Theoreticians try to piece together a coherent model of the formation of the planets, and to explain the motions of Saturn's rings. Geophysicists attempt to fit a handful of rocks from the Moon into a history of the Solar System that is consistent with observations of other planets' surfaces. Biologists are left to ponder perhaps the greatest mystery of all: why the Earth is the only place in the Solar System where life has evolved.

Taken together, these diverse lines of enquiry have given us a fuller understanding of where we come from and how our planet has evolved. And this most enthralling scientific detective story of modern times has allowed us, for the first time ever, to gain a true sense of our place in the cosmic scheme of things.

▲ **Early Moon-shot** *One of the first close-up pictures of the Moon, taken from the US Ranger 7 probe in July 1964 from an altitude of 60 km, shortly before the spacecraft made its pre-programmed crash-landing on the Moon.*

of 1957–58 gave the superpowers a target to aim for: to launch the first satellites and claim political mastery of the heavens. Thankfully, the thrust of scientific discovery has outlasted the enmities and political brinkmanship of the Cold War.

▶ **Flag-waving exercise?** *The Apollo Moon landings in the late 1960s and early 1970s were the result of a direct political goal, and some have claimed that the true scientific benefits were small. Even so, during this period money was made freely available for less politically attractive missions exploring the Solar System with uncrewed vehicles.*

Hand in hand with this scientific revolution has been an equally far-reaching revolution in engineering. To turn elusive dreams of exploration into reality required engineering of unprecedented precision. Crewless spacecraft had to be able to operate automatically for many years with only minimal intervention from controllers on Earth. To send such craft into space required the skills of engineers to navigate through space using the time-honoured method of tracking star positions and to orchestrate rocket propulsion to the finest limits, taking celestial mechanics to the cutting edge of scientific research. In time, engineers and scientists have taken the laws of celestial mechanics to the fore-front of research, and learned how to "steal" gravitational energy from one planet in order to "slingshot" spacecraft on to other worlds. They have built and operated a network of ground stations capable of monitoring the extremely faint signals from distant spacecraft. They have learned to work around severe handicaps in communication, such as the finite speed of light which makes real-time relays in space an impossibility. And with advanced technology they have overcome problems that arose at the most inopportune of moments.

The rise of JPL

Space travel has not always been viewed with awe. In the early years of this century, travel to the planets was regarded as the preserve of a lunatic fringe who aroused only derision from the more established sections of the scientific community. When a group of enthusiasts and graduate students at the California Institute of Technology (Caltech) began to test rockets in the mid-1930s, they termed their research establishment the Jet Propulsion Laboratory (JPL) to give it a veneer of respectability.

After the Second World War, US rocket research was bolstered by the arrival of a group of German émigrés, led by the redoubtable Wernher von Braun, whose earlier attempts to reach space, as cynics gleefully pointed out, had only helped to lay waste London. Yet the technical heritage of the German V-2 rocket, upgraded with designs for ever more powerful engines, enabled the United States to enter the space age and later create an agency that became a byword for technical excellence.

When the National Aeronautics and Space Administration (NASA) was formed in 1958, only

◄ **The first frontier** The 1990s have seen a minor resurgence of interest in the Moon. Largely forgotten during the rush to explore the planets, the Moon has still not been fully surveyed. The prominent crater Tycho is seen here, in an image taken with the Ultraviolet/ Visible Camera aboard the Clementine probe in 1994.

◀ **Jet Propulsion Laboratory, Pasadena** When a group of Caltech graduates began rocket experiments in 1944, they had little idea that within 20 years their organization would become the world's leading centre for planetary exploration. Today JPL employs nearly 5,000 people, some of whom work for the Deep Space Network, which continually monitors signals from US planetary spacecraft in all parts of the Solar System.

▶ **European Centre** The Operations Centre of the European Space Agency (ESA) at Darmstadt, Germany. ESA was formed in 1975 to co-ordinate the efforts in space research and technology of 13 European countries. ESA's biggest success to date has been the Giotto space probe mission to Halley's Comet. Giotto passed within 600 km of the comet on the night of 13–14 March 1986. The control room is pictured here at the time of Giotto's Halley fly-by.

JPL had experience of building and launching spacecraft. Because it was operated by Caltech and contracted to NASA, JPL was an outsider and its scientists wanted little part in the plan to launch astronauts into space for reasons that had more to do with political prestige than scientific enquiry. Instead, they provided NASA with a blueprint outlining how to explore the planets with JPL-built spacecraft. The self-confidence of the Kennedy era soon manifested itself in the all-encompassing national goal of sending spacecraft, and ultimately humans, to the Moon. As a result of the accompanying surge in funding, JPL was able to accomplish its aim of sending missions to other worlds, beginning with Mariner 2's successful fly-by of Venus in December 1962. So began the golden age of planetary exploration, which drew to a close in 1989 when Voyager 2 finally reached Neptune.

▲ Voyager I lift-off
The Voyager missions were without doubt the most successful endeavour of JPL and, perhaps, of planetary exploration as a whole. Here Voyager I blasts off on 5 September 1977, 16 days after Voyager 2.

produced the Pioneer craft. These lived up to their names by landing on the surface of Venus and making the first fly-bys of Jupiter and Saturn.

The European Space Agency (ESA) made its dramatic entry onto the interplanetary stage with the Giotto probe, which in March 1986 made a close pass of Halley's Comet and went on to examine Comet Grigg-Skjellerup in July 1992. ESA scientists now plan to send a more sophisticated probe to sample the surface of another comet at the turn of the century. ESA will also attempt to land a craft on the cloud-shrouded surface of Saturn's moon Titan at roughly the same time.

In a Moscow no longer cloaked in secrecy, the Space Research Institute is now trying to forge its position in the unstable Russian economy with an ambitious series of missions to Mars.

The Japanese NASDA and ISAS agencies joined the club with fly-bys of Halley's Comet in 1986 by the Suisei and Sakigake probes; and in the 1990s Japan's Yohkoh solar observatory began returning dazzling X-ray photographs of the Sun.

Despite the media attention paid to these missions, most of the people who have worked on them remain unknown to the general public. Their obscurity is symptomatic of the criticism most often levelled at planetary exploration: that robots are not as interesting as humans, and consequently uncrewed missions are less dramatic than those involving astronauts. Yet the story outlined in these pages is one of extraordinary human endeavour, rich in drama and surprise, and there are some very human stories behind the robotic triumphs and impersonal photographs. Behind every planetary photograph is a compelling story of ingenuity and and an enduring testimony to the triumph of the indomitable human spirit.

Future imperfect

The most recent spacecraft sent to other worlds bear as much resemblance to the early Mariners as a Lamborghini does to a Model T Ford. Partly as a result of the race to the Moon in the 1960s, technical advance in miniaturization has streamlined our robotic envoys and at the same time enabled them to carry out more detailed scientific investigations. But the flipside of this increasing complexity has been that each new mission has become vulnerable to heart-breaking failures, the consequence of ever more improbable technical mishaps.

The early 1990s were disastrous years for space science generally, and for planetary exploration in particular. The undoubted low point came in August 1993 when, just hours before it was due to enter Mars orbit, all contact with the Mars Observer probe was lost.

In the process, JPL blossomed into a self-contained science city in the San Gabriel foothills, north of the Rose Bowl in Pasadena. It is still the world's leading centre for planetary research, its scientists constantly planning new and more innovative ways of exploring the planets. But it is no longer alone; other centres and agencies around the world have joined in. South of San Francisco, for instance, is NASA's Ames Research Center, which

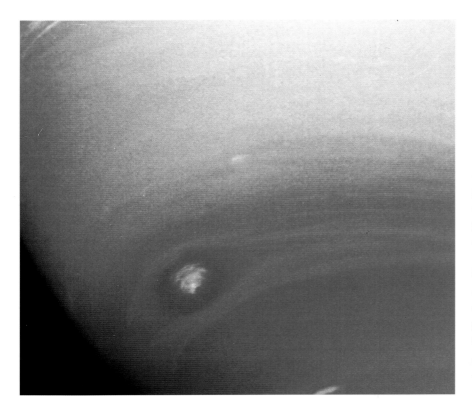

◀ **Neptune's Great Dark Spot** *Voyager 2 glimpsed this splendid view of Neptune, during its final planetary encounter in August 1989.*

▶ **Repair mission** *In December 1993, the crew of space shuttle* Endeavour *put the finishing touches to the refit and repair of the Hubble Space Telescope, 570 kilometres above Western Australia.*

The success of JPL in steering a legion of probes through the Solar System made the failure of Mars Observer all the more stark — it was the US space programme's first catastrophic failure of a planetary mission since Surveyor 4 crashed onto the Moon in 1967. Yet in the early 1960s, when the early Ranger missions to the Moon had repeatedly failed en route, soothsayers had predicted the demise of JPL and an end to planetary exploration. This did not happen.

Today, the price of failure, in both financial and political terms, is higher than ever before. NASA itself is only too painfully aware of this fact and is striving for what it calls "faster, better, cheaper" missions. In the late 1990s a new generation of smaller probes, known as the Discovery spacecraft, will be launched to targets within the inner Solar System at a fraction of the cost of current missions — the Mars Observer nudged beyond the $600 million mark, while the Galileo probe, now en route to Jupiter, came to nearer $1,000 million.

Galileo has come, in many ways, to symbolize the trials and tribulations of planetary exploration in the 1990s. It suffered a serious technical fault that threatened its ability to return data to Earth — the main antenna was prevented from opening fully by an obstruction perhaps a few millimetres across. But, as they have done many times before, JPL

engineers worked around the problem and salvaged a high-profile mission that was intended to be NASA's "flagship planetary explorer for the 1990s".

How the next steps in planetary exploration will be taken in the New World Order is difficult to discern, for the twin vexations of political and financial uncertainty continue to plague most endeavours in space. With the ending of the Cold War, it is commonly said that many of the old reasons for space exploration no longer exist. Planetary exploration has always been forced to eke out a Cinderella-like existence, the poor relation of the more glamourous and expensive programmes such as space shuttles and space stations. At a time when many question

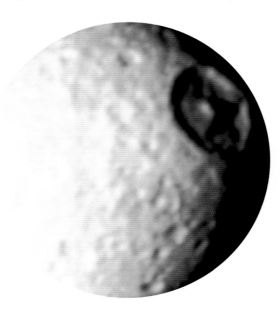

▶ **Mimas** *Although poorly imaged by Voyager, the 130-km crater Herschel clearly dominates Mimas,* the innermost moon of Saturn. Astronomers are surprised Mimas was not torn apart by the impact.

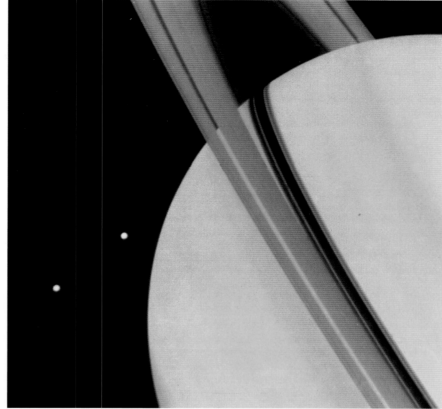

the need to venture into space at all, it is worth remembering that NASA's annual budget for planetary exploration has for many years been no more than the equivalent of an evening at the movies for every American citizen. But the images returned — for which nature provided the special effects — are more than mere entertainment.

Now is the time to review the state of knowledge about the Solar System. This highly illustrated volume is not an astronomy book in the conventional sense, for it does not tell you how to find and observe the planets. It simply attempts to chronicle the important scientific discoveries of over three decades of planetary exploration, to place them in historical perspective, and to imbue them with a sprinkling of anecdote. There are, within these pages, some remarkable stories and many even more astounding images from what has been, without any possible doubt, the most exciting scientific adventure of all time.

▲ **Saturn** The ringed planet, as viewed by Voyager 1 in November 1980 from 13 million km. The Cassini Division in the rings can be clearly seen, along with the shadow cast by the rings on the surface of Saturn. The moons Tethys and Dione are visible on the left of the image.

▶ **Europa** One of Jupiter's weird moons, seen in detail for the first time by the Voyager missions. The network of grooves on its icy surface is unexplained even today.

LIES, DAMNED LIES AND IMAGE PROCESSING

Hand in hand with the revolution in our understanding of the planets, there has been a far more subtle change in the way that information about them is presented. The information-technology revolution has transformed our ability to visualize the planets and their moons and to present data in a form that can be more readily digested. Books on planetary exploration usually skate over these developments, but the ways in which spacecraft take pictures are so fundamental to the progress made that the subject will be examined here in some detail.

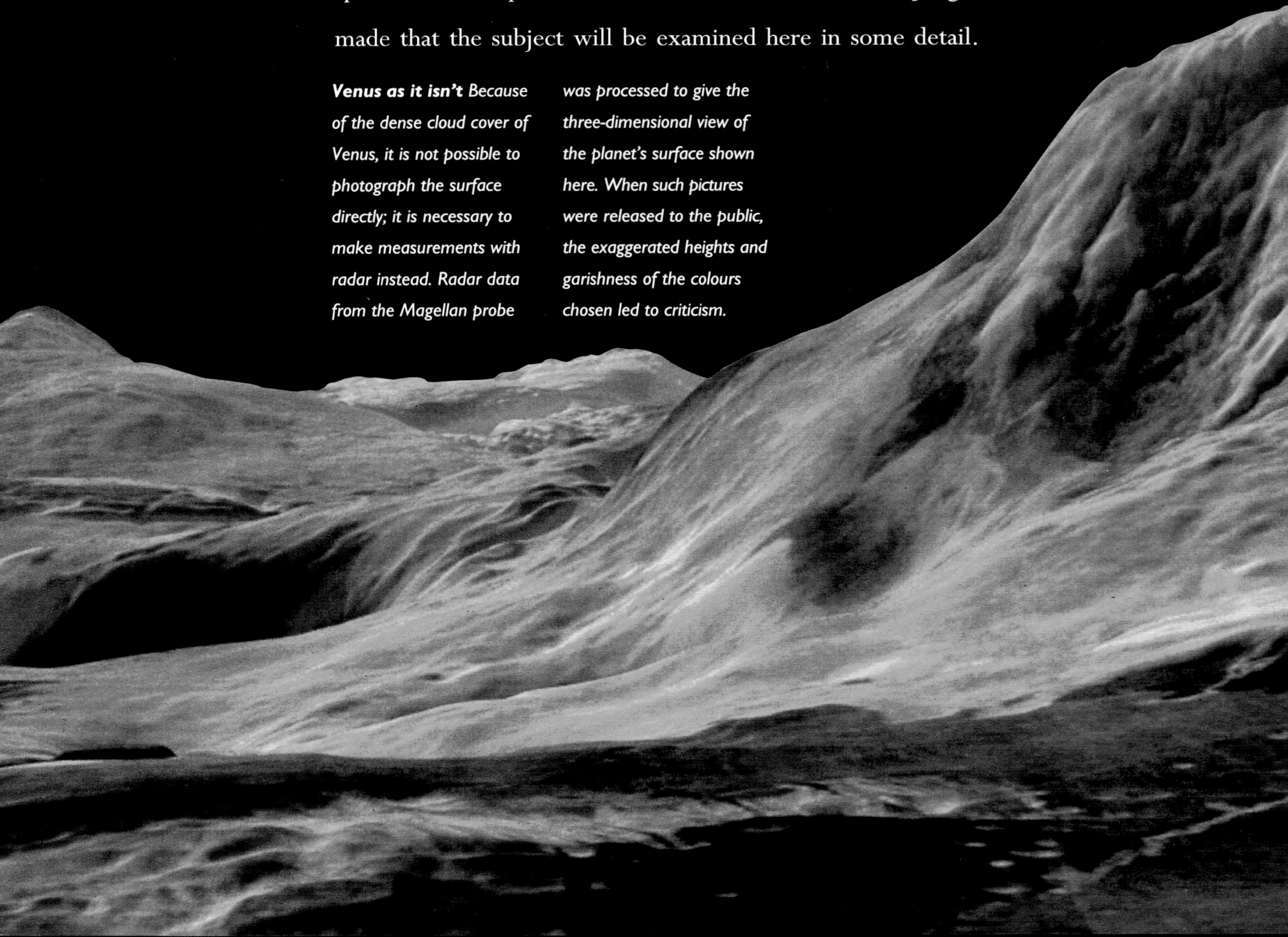

Venus as it isn't Because of the dense cloud cover of Venus, it is not possible to photograph the surface directly; it is necessary to make measurements with radar instead. Radar data from the Magellan probe was processed to give the three-dimensional view of the planet's surface shown here. When such pictures were released to the public, the exaggerated heights and garishness of the colours chosen led to criticism.

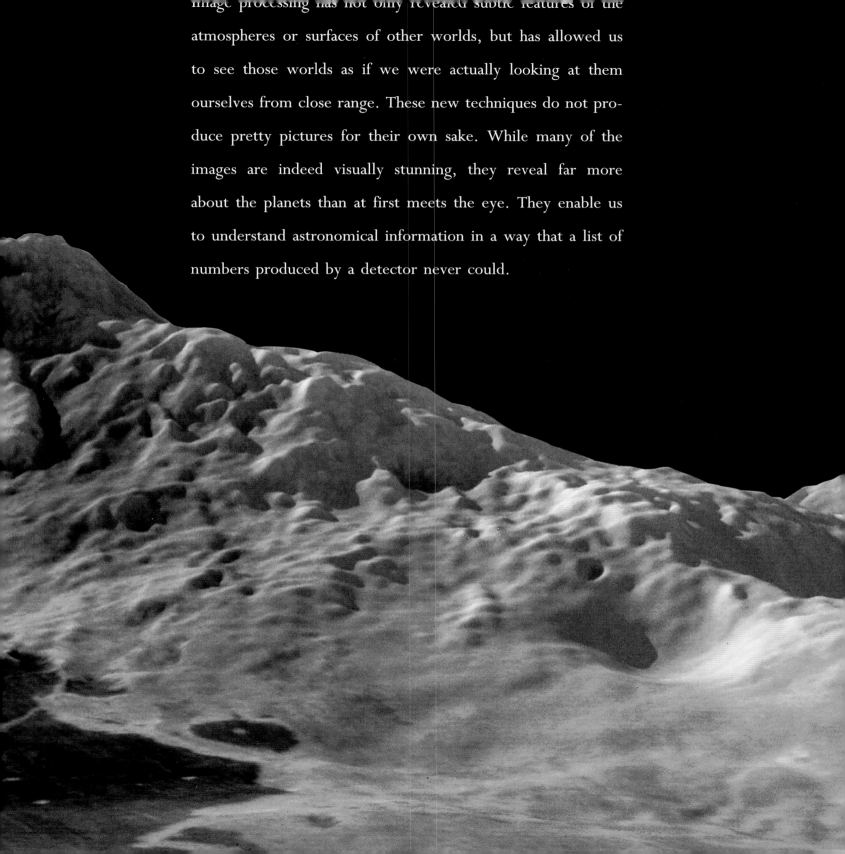

image processing has not only revealed subtle features of the atmospheres or surfaces of other worlds, but has allowed us to see those worlds as if we were actually looking at them ourselves from close range. These new techniques do not produce pretty pictures for their own sake. While many of the images are indeed visually stunning, they reveal far more about the planets than at first meets the eye. They enable us to understand astronomical information in a way that a list of numbers produced by a detector never could.

◄▼ **A trick of the light**
The full Moon is a familiar sight (below) and appears bright in the night sky. The lunar maria ("seas") are darker regions and for centuries were mistaken for oceans. Brighter, rayed craters are also visible in the view, which was obtained by the Apollo 11 astronauts. Yet when they landed on the surface, the Apollo astronauts found the whole of the surface to be very dark. The spectacular view of the Taurus-Littrow valley (left) was taken by the last astronauts to walk on the Moon during Apollo 17.

To begin, we must consider the eye itself. When the German scientist Hermann von Helmholtz determined how the human eye worked in the nineteenth century, he remarked that if he were supplied with an optical instrument of similar construction he would refuse to accept it because it was so poorly designed. It is very easy to trick the eye into seeing illusions. Helmholtz determined that the retina is lined with cone-shaped cells of three types, each sensitive to a different spread of wavelengths of visible light. It is only when signals from the three cone-types are combined in the brain that we are able to sense a full colour image.

Astronomically speaking, optical illusions caused by tricks played on us by the eye are common. Consider the full Moon. In a cloud-free night sky the Moon appears very bright, yet when the Apollo astronauts landed on the Moon the surface appeared to be dark. In truth, Moon rock is indeed dark, absorbing some 93 per cent of the sunlight reaching it; but compared with the extreme blackness of space, the Moon seems bright. If we could hang a giant white sheet behind the Moon, we would see just how dark our celestial neighbour really is.

The problem of depicting true colours has caused a number of dilemmas for optical scientists. Perhaps the most controversial instance in the past few years was the strange case of the volcanoes of Venus.

Colour me badly

One of the more recent dividends of planetary exploration came in the form of startling images of the surface of Venus from the Magellan spacecraft. Because Venus is perpetually enshrouded in dense clouds, a powerful radar system was used to "see" the surface from orbit. By accurately matching indi-

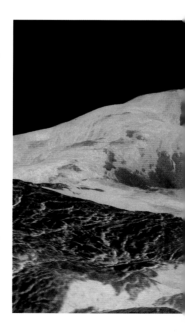

vidual points on different images, it was possible to generate three-dimensional views of the Venusian surface. Traditionally, the matching of image points was a laborious business, but now computers are able to complete the task in a few minutes.

When NASA released the first Magellan pictures processed in this manner, the surface of Venus was portrayed as an electric-orange, vertiginous landscape beneath a pitch black sky, which in reality it most assuredly is not. Immediately the agency came under heavy criticism for having exaggerated vertical scales and enhanced features to make them look like active lava flows; there were fears that non-technical viewers were being misled. In fact, the exaggeration of vertical scales is a fairly common practice in photogrammetry (the science of making measurements from photographic information), but this was not explained to the public at the time. NASA listened, and a less exaggerated view of one particular volcano was later released.

The correct depiction of the surface colour of Venus was more complicated because the Magellan observations were made using radar, not visible light. Scientists at the Jet Propulsion Laboratory took their cue from the Soviet Venera 13 and 14 spacecraft, which landed on the surface in 1982. The probes found that the surface rocks appear to be yellow, and chemical analysis revealed them to be of a basaltic type. Basaltic rock, as found in lava on Earth in places like Hawaii, has a greyish-yellow hue, so scientists took these tones and altered the processed images from Magellan accordingly.

The truth is that nearly all images returned from space are enhanced to some extent, and few of them show exactly what a human stowaway would see from the same location. But the original raw data from spacecraft are not usually in a form that can be readily understood by a human observer and, in order for them to have scientific value, they are quite often coloured in an unrealistic manner.

▼▶ Maat Mons *The largest volcano on Venus, Maat Mons, rises some 8 km above the surrounding terrain. When NASA made the image below in October 1991, they exaggerated the vertical scale more than 22 times. Later (right) this was reduced to ten times. The surface coloration is entirely invented; a more realistic interpretation, based on the colours of basaltic rock as seen by Venera 13, is shown on pages 16–17.*

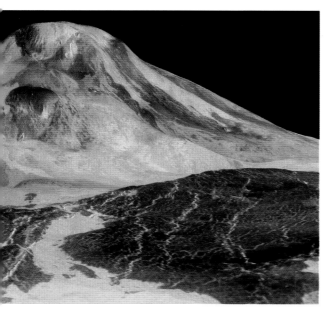

▶ Venera 13 *On 1 March 1982, the Soviet Venera 13 lander successfully reached the surface of Venus and survived for 127 minutes, enough time to record this image. Colour was obtained by taking three separate scans through red, green and blue filters; these were combined into a full-colour image with the aid of the colour scale visible in the foreground. Seen here is a small portion of the image, which appears slightly curved because of the way the camera scanned.*

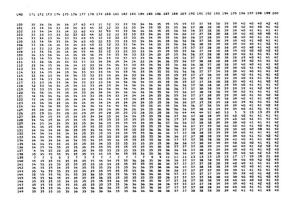

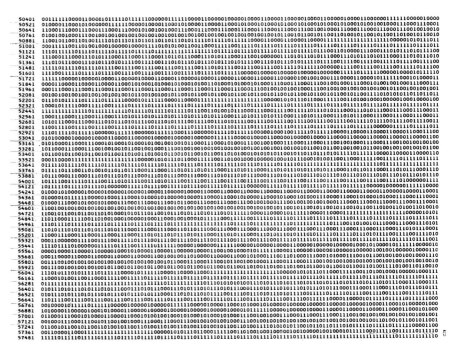

▲▶ Zeros and ones *Pictures from spacecraft are transmitted as a series of radio signals representing zeros and ones (binary digits or "bits"). In this early example (right), each group* of six zeros and ones gives a number from 0 to 63 (above), each corresponding to a shade of grey. The complete image is then assembled from these shaded dots by computer.

Pixel pushers

When the first spacecraft were launched to the Moon in the late 1950s, the technology available for transmitting visual images was fairly crude. It was found necessary to adapt a full photographic laboratory, place it inside the craft, scan the photographs it produced and relay the signals back to Earth. That, in essence, was how the imaging systems on the US Lunar Orbiter series of probes worked. Their task in the mid to late 1960s was to map the whole of the Moon and, in the process, reconnoitre landing sites for the Apollo astronauts. The system they used was originally developed by Eastman Kodak for the first US military spy satellites. The film was developed on board, the negative scanned by a miniature television camera and the output transmitted to Earth, where a replica photograph was assembled. This was cumbersome, but it worked.

The first time television technology was used directly was on NASA's Ranger spacecraft, whose task was to take close-up pictures of the Moon before crash-landing on its surface. The technology used was the vidicon tube, similar to the tube in an ordinary television set. Power restrictions meant that a complete television picture, composed of many hundreds of individual lines, could not be transmitted back to Earth in one go. Instead, pictures were sent one line at a time, as a sequence of brightness values for all the points in the line. Back on Earth, these values were used to expose film line by line. Transmission difficulties meant there were often lines missing; and interference and weakness of the signal caused further difficulties. The optical engineers at JPL had to learn to circumvent these technical problems.

▶ Mars digitized *In July 1965, Mariner 4 returned the first close-ups of Mars. They were transmitted in binary digit form, giving a grainy, "pixelated" texture to the images. Because of the craft's limited power, it took ten days to send all the Mariner 4 observations to Earth as digital signals. This picture was the ninth individual frame from a total of only 22 taken, just ten of which actually showed surface features.*

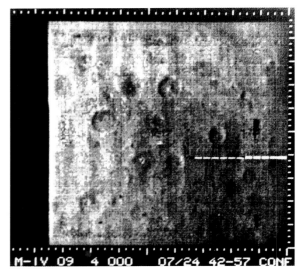

Planetary exploration gave the English language the word "pixel" (meaning picture element) and led to the refining of an imaging technique now standard in many fields of science. Spacecraft vidicons split an image into hundreds of thousands of tiny squares – the picture elements. For each square the vidicon sensor records a brightness value as a number, usually between 0 and 255, each number corresponding to a shade of grey from intense white down to pitch black. The image is then sent back to Earth as a stream of numbers and reconstructed in reverse. It is, in effect, just a Space Age version of painting by numbers, and has gained the JPL optical engineers the sobriquet of "pixel pushers".

Coloured pictures are obtained by a slightly modified technique. The image is recorded as shades of grey as before, but in three parts. Each image is obtained through one of three differently

▶ **Reds, greens and blues** *In 1991, the Hubble Space Telescope returned colour pictures of Mars with its Wide-Field Planetary Camera. As shown here, each colour image was reconstructed from three separate pictures, each taken through a different filter, coloured red, green or blue. This three-separation technique is standard practice in space-based imaging systems.*

coloured filters – usually the primary colours of red, green and blue. An example is the image of Mars obtained by the Hubble Space Telescope (shown above). Three separate images were taken in different single colours and then combined into one to produce a reasonably realistic full-colour image.

There are, however, limitations to this technique, which are set by the speed at which the spacecraft can transmit data and by the sensitivity of the vidicon – or its successor, the CCD (of which more later). Both the transmission rate and scanner sensitivity have improved immensely in recent years, but the same basic technique is still used to obtain colour images from space probes at the outer limits of the Solar System.

Shades of pink and blue

Mars – the Red Planet – is the only planet visible from Earth with the naked eye to have a distinct and instantly identifiable colour. When NASA's two Viking spacecraft landed on the planet in 1976 their cameras revealed a barren desert landscape strewn with rocks. The surface colour of the soil turned out, as expected, to be a rich rusty brown; but the real surprise was the colour of the Martian sky.

Observations from space had already shown that Mars has a thin, but meteorologically active, atmosphere containing a great deal of airborne dust. The

Viking lander cameras scanned horizontally very slowly and recorded images through a number of filters. Each lander had a colour-coded "test card" against which these observations could be gauged.

When the first eagerly awaited colour images were received from Viking 1 and reassembled on Earth, they were "surprising and disquieting" in the words of the camera team leader, the late Thomas ("Tim") Mutch. Recording his thoughts in the official NASA history of the Viking camera, *The Martian Landscape*, he recalled: "The entire scene, ground and atmosphere alike, was bathed in a reddish glow. Unwilling to commit ourselves publicly to this provocative display, we adjusted the parameters in the calibration program until the sky came out a neutral grey." When this image was released to the press, photographic processing had inadvertently given a blue tinge to the sky. Reporters quickly picked up on this blue colour, and remarked that the Martian sky was comfortingly Earth-like. A few hours later, while looking at the published photographs, one of the imaging-team scientists noticed the colour of a wire protruding from the body of the Viking lander did not look quite right. When the image colour was adjusted, the scene returned to how it had originally been seen, with a salmon-pink sky – a colour caused by the scattering of light by dust in the Martian atmosphere.

▶ **Uranus enhanced** *Of the four planets it visited, Voyager 2 found Uranus the least colourful, and extreme exaggeration of colour tones was needed to bring out details on the surface. The false orange-coloration in the pole (at right) probably results from a thin haze of hydrocarbon molecules high up in the Uranian atmosphere.*

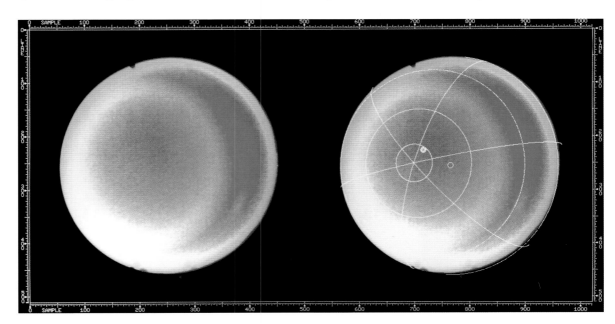

◄ *Martian chronicle*
Throughout the course of a Martian year, the brightness of the sky as seen by the Viking I lander changed little. Although fogs and clouds could often be seen from orbit, they were not detected by the landers. These images, in which the sky colour has been altered without changing the overall brightness, were obtained in 1980-81.

Since that rather illuminating episode, images of Mars have been released in two basic forms: either as the scene would appear if you were actually standing on Mars yourself, or how the scene would appear if illuminated with the light conditions on Earth. The "Earth lit" view has its special benefits: it enables geologists to use familiar terrestrial techniques when looking for differences in the surface geology. For example, scientists at Washington University in St. Louis produced images in the late 1980s that gave Mars a curiously azure-coloured sky; it was this lighting that best brought out interesting features in the surface rocks and differences in their mineral composition.

Real or unreal?

Thousands of remarkably colourful images of the gas giant worlds, Jupiter, Saturn, Uranus and Neptune, were returned by the two Voyager probes. On the whole, colours in the Voyager images were greatly enhanced — sometimes almost to the point of garishness — to reveal features that would otherwise have been difficult to discern. In fact, even the unenhanced Voyager images were in "false" colour — the result of a quirk of technology.

Both of the Voyager probes carried a vidicon-type imaging system, which included both a wide-angle camera and a narrow-angle camera. Colour images were obtained from these by using the usual

▲▶ **Red Planet blues**
How would we see the Martian landscape if we could stand on the surface? The best approximations are shown in these images processed from the original Viking photographs. Dust in the Martian atmosphere gives the sky a distinctly pink colour. Realistic views were obtained only thanks to intimate knowledge of the camera's spectral response and by colour-matching with the "test chart" on the lander (see image at right). But geologists have found they can best identify surface minerals on Mars by replacing the pink sky with a blue one, to approximate lighting conditions on Earth.

technique of taking three black-and-white shots through different filters and then combining these to give a full-colour picture. But for Voyager, instead of the standard red, green and blue filters, new colours were often used.

Because of the way that the camera worked, each Voyager image took up to 48 seconds to record. However, for the semiconductor materials used in the Voyager vidicons to store red light for more than a few seconds, they would have needed to be cooled, adding to the cost and complexity of the mission. So, in place of a red filter, Voyager made do with orange.

Two other filter colours were also employed by Voyager. Because of the importance of methane and sodium in the outer Solar System, filters were used that select light at the wavelengths sodium and methane emit

◄ Psychedelic hues
Jupiter's volcanically active moon Io is shown in false colours in this enhanced image returned by Voyager I in March 1979. Orange regions are composed of sulphur-rich lava, while blue areas reveal a "snow" of sulphur dioxide. Compare the colours in this image with pictures of Io shown elsewhere in this book.

▼ Sulphur halo Even from Earth-orbit the signs of Io's volcanic activity can be detected. In early 1994, the Extreme Ultraviolet Explorer returned this "pixelated" view of ionized sulphur around the Jovian moon.

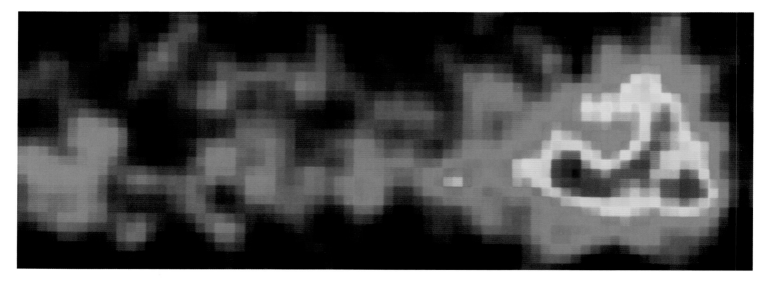

most strongly. When images taken through such filters are recombined, they inevitably reveal features that cannot be seen as well by the human eye, and introduce "false" hues to the image.

Thus, although the Voyager pictures are not "true" colour reproductions, they provide scientists with useful information. Take, for example, the images of one of Jupiter's moons, Io, the surface of which is dominated by sulphur compounds emitted by volcanoes. As anyone who has heated sulphur in a school laboratory knows, it changes from yellow to orange to red to black. Io's surface is coloured in the same way, and these colour differences enable geologists to determine exactly how Io's surface has changed with successive eruptions. Most of the published photographs of Io, have been almost over-enhanced

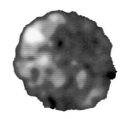

◄ No change on Io
In the spring of 1992, the Faint Object Camera of the unrepaired Hubble Space Telescope recorded the upper of these two images of Jupiter's moon Io. Although no features less than 250 km across could be distinguished, this observation showed that the surface of Io had not changed appreciably since the Voyager encounters in 1979, when the lower image was produced.

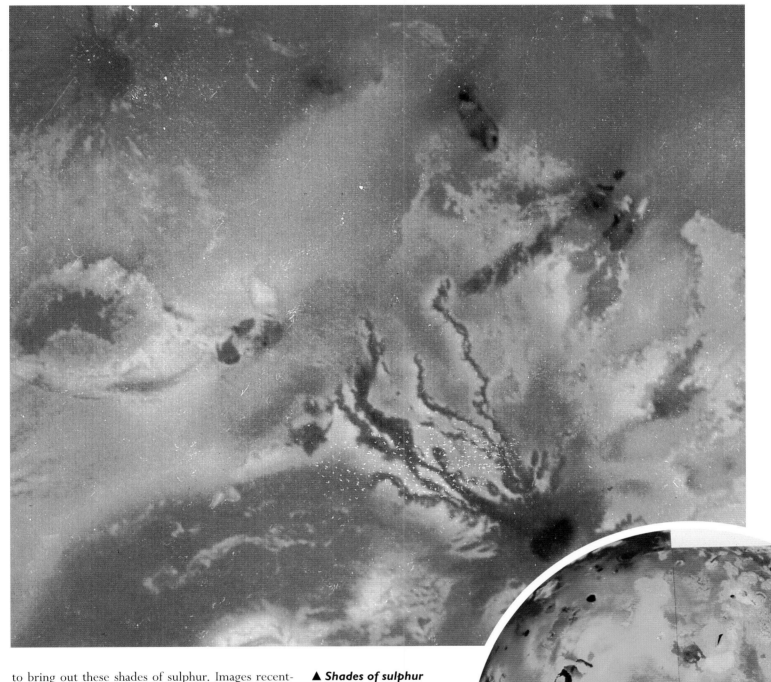

to bring out these shades of sulphur. Images recently released show far more subtle hues on Io, with muted pastel colours, including greens.

The Galileo spacecraft, which will reach Jupiter in late 1995, was equipped to give more realistic views with its CCD (charge-coupled device) camera. CCD cameras contain a silicon surface, which is broken up into a matrix of small picture-elements, each able to detect the impact of the individual photons of light. Because of this extreme sensitivity, the CCD can detect greater ranges of brightness than the Voyagers' vidicons, and can assemble an image in thousandths of a second, without the need for exotic filters. Galileo's CCD camera will provide the most "realistic" observations of Jupiter and its moons so far obtained.

▲ **Shades of sulphur**
Lava flows stretch for hundreds of kilometres from Ra Patera on Io. The colours are exaggerated to show clearly sulphur at various stages of heating. Impurities can alter these hues further.

▶ **Celestial Pizza** *Frames taken by Voyager 1 were assembled to produce this false-colour photomosaic of Io. The hoofprint-shaped feature at lower right is Io's largest volcano, Pele.*

THE REALM OF THE SUN

The dominant object in our Solar System is the star at its centre.
The Sun is a giant ball of luminous gas, which shines because
its hydrogen is being converted to helium by the process of
thermonuclear fusion. At 150 million kilometres distant, it
is the star next door. Fairly average as far as stars
go (even a little on the small size), the
Sun is remarkable only for its
proximity to Earth and as
the point to which the
Earth is gravitational-
ly bound. But with-
out the Sun, ad-
vanced forms of
life would never
have evolved on

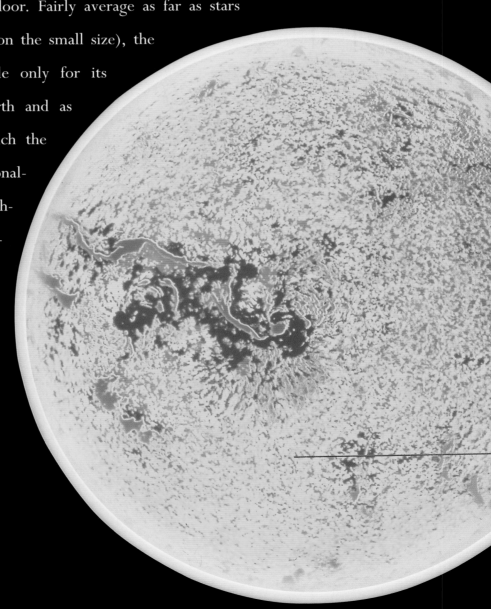

Earth, because the Sun provides all of the heat and light that is needed for all but the very simplest of life-forms to exist.

The Sun holds the key to the origins and ultimate fate of our planetary system. Astronomers know that the life cycles of stars take in many thousands of millions of years. The Sun is roughly halfway through its life-span and, within the next 4,500 million years, it will slowly expand to become a red giant when its prodigious consumption of hydrogen exhausts its available reserves. Deep within all stars, light elements (hydrogen and helium) are fused into heavier ones in the thermonuclear furnaces that lie at their cores. Often a star's life ends in a supernova, a spectacular explosion that creates the heaviest chemical elements and spews them into space. These heavy elements, in turn, provide the building blocks of planetary genesis in the form of dust and gas.

Polar explorer An artist's impression of the European Space Agency's Ulysses probe approaching the Sun. In the summer of 1994, Ulysses became the first artificial object to pass over the south pole of the Sun, and in autumn 1995 it passed over the north pole. Once all of the space probe's data has been fully analysed, astronomers will know far more about the part that the Sun plays in generating the high-energy particles that flood the Solar System.

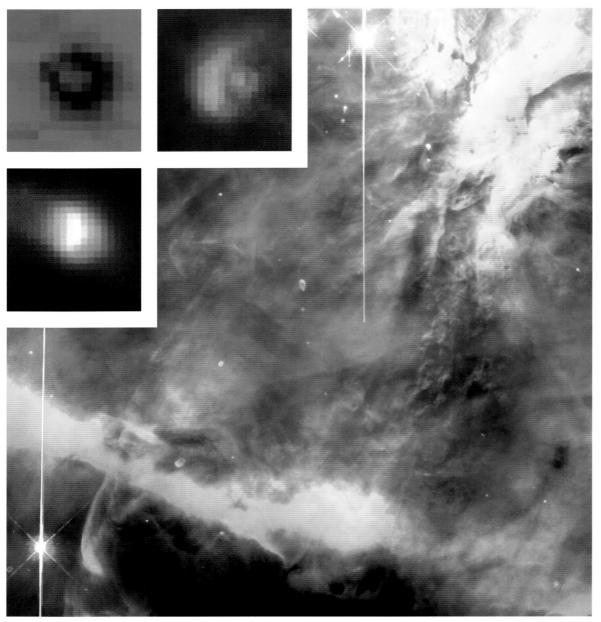

◄ **Heat and dust** Stars and planets are born in the luminous clouds of dust and gas that litter our galaxy. The main image, returned by the newly refurbished Hubble Space Telescope in January 1994, shows the most famous of these "stellar nurseries", the Orion Nebula, M42. It is easily visible with the naked eye, in the "sword" of the famous constellation. Young, bright stars are visible within the luminous gas and dust. Many astronomers believe that the Orion Nebula contains a number of proto-solar systems. The insets show enlarged views of such systems, each roughly twice the size of our own, taken by the Hubble telescope before its optics were refurbished. Each disk has a cool star at the centre, and the disk itself contains planets in the making.

Some 4,500 million years ago, a conglomeration of such dusty material in our part of the Milky Way galaxy – out towards its periphery – started to collapse in on itself. It did so under its own gravity, perhaps helped along by compression from the shock wave caused by a nearby supernova explosion. The result was a swirling region of dust and gas, whose centre grew to be so dense and hot that it underwent thermonuclear ignition. The Sun was born, and the stream of radiation it emitted was to shape much of what immediately followed.

Planetary origins

The full story of how the Solar System came to be the way it is remains to be established, and much of what follows is necessarily conjectural. Many of the facts have been discovered only thanks to the craft sent into space in the last few decades. We can

hope that many of the remaining mysteries will be solved by future missions. As far as the broader picture is concerned there is a consensus among most planetary scientists, upon which the following explanation is based.

The gaseous dusty material that surrounded the newly born Sun gradually spun itself into a flattened elongated disk. The disk was unstable, so much so that eddies formed within it, and as a result innumerable clumps of dusty material aggregated. These so-called "planetesimals", each perhaps a few kilometres across, formed the building blocks for the planets of the Solar System.

The chemical nature of these planetesimals depended largely on the distance they were formed from the Sun. Thanks to the heating from the Sun, material in the planetary disk closer in to the Sun remained at a higher temperature than that further out. Close in, only rocky materials such as oxides of aluminium and silicon were able to condense at the high temperatures. More volatile materials — those that condense only at lower temperatures, such as compounds containing sulphur or water — were blown out of the inner zones by the solar wind and formed only further out. So, as planetesimals aggregated to form larger bodies, their chemical composition was dictated largely by their distance from the Sun. Close in, the proto-planets were composed mostly of silicate rocks; while those further out were larger and gaseous. This explains why the planets of the Solar System seem to fall into two classes. The inner planets are relatively small, rocky worlds, Earth-like in nature, and so are usually termed terrestrial planets. The outer planets are huge, gaseous worlds orbited by ring systems and with large retinues of moons.

This process of planetary formation was a messy one. A thousand million years or so after the planets began to form, there were still innumerable planetesimals roaming the Solar System. Many were attracted by the gravitational pulls of the larger worlds, and crashed into their surfaces creating huge craters. Those worlds that have evolved little since that cratering period still have these craters as a record of the earliest epochs of the Solar System. Larger planetesimals seem to have caused greater damage when they crashed into the proto-planets, penetrating deep below their crusts. One such impact may have excavated the Moon from the Earth; another blasted into space most of the proto-planet that later became Mercury, and pushed it into an eccentric orbit closer to the Sun.

The diversity of the planets today reveals that not all of them developed in the same way. The size of a planet was the key factor. Planets are heated in a small number of basic ways. When they form, energy is released due to gravitational compression and the impact of planetesimals on the surface. The outer part of the Moon, for example, may well have been completely molten some 4,000 million years ago as a result of such processes. More importantly, planets are heated by the natural radioactive decay of materials within them, and the extent of this radioactive heating depends on their size. Large bodies retain proportionately greater amounts of this energy because they have a higher ratio of bulk to surface area and so the energy finds it more difficult to escape. In large planets, radioactive heating eventually melts the interior. In this liquid state, denser elements form a mantle and crust through a process called differentiation.

By the time that the process of planetary formation was essentially complete, there was still some material left over. Separating the terrestrial planets from the gas giants is a region of interplanetary rubble: the asteroids. It is generally believed that the asteroid belt contains material that was prevented from coalescing into a single body by the immense gravitational attraction of Jupiter. Part of this asteroidal material occasionally reaches the Earth as meteorites. Comets, too, hold important clues to our planetary origins because they contain pristine, icy material, which has effectively been in "deep freeze" since the formation of the planets. Comets are only spectacular in the sky when they pass close to the Sun and their surfaces are heated, which vaporizes material that streams out to form a bright tail. It has been realized in recent years that asteroids and comets have more in common with each other than had at first been suspected, and both undoubtedly hold important clues about the early history of the Solar System.

▼ **Planetary formation**
An artist's impression of the birth of the Solar System some 4,500 million years ago. The Sun at the centre has undergone thermonuclear ignition, and the clumps of material shown in the foreground are the "planetesimals" from which planets will form.

Bright star

There are, to be sure, many unsolved riddles about the star around which we orbit, but the basic facts are certain. The Sun is just one of many thousands of millions of stars, and both astronomy textbooks and our own experience suggest that it is reasonably stable. The Sun contains an estimated 99.4 per cent of all the material in the Solar System, has a radius of just over 696,000 kilometres and rotates about its own axis once every 27 days. But its bright surface, the photosphere, is a barrier that prevents us from easily learning more about the Sun, because it shields the interior from direct observation. The brilliance of the photosphere also hides the full extent of the Sun's outer atmosphere from view.

Since the 1970s, probes dedicated to solar observation and telescopes aboard other spacecraft have revealed fascinating new information about the Sun. In fact, the visible radiation that emanates from the Sun is just over 40 per cent of its total energy output. To get a fuller picture, it is necessary to go into space to observe the highest energy radiations – ultraviolet, X-rays, and gamma-rays – which are absorbed by the Earth's atmosphere.

At its centre, the Sun is believed to have a temperature of about 15 million degrees C, and a pressure 300,000 million times greater than that at the Earth's surface. These extreme conditions act to fuse atoms of hydrogen into helium in an immensely powerful nuclear furnace. Energy is transferred outwards mainly in the form of X-rays and gamma-rays, the most energetic forms of electromagnetic radiation. In the "radiative" zone surrounding the core, the pressure is still tremendous and the density of material is, on average, millions of times greater than in our own atmosphere. The energy that radiates outwards is continually being absorbed and re-absorbed by the molecules in this region, and it may take ten million years for it to trickle outwards to the Sun's surface and then be radiated into space. When this energy reaches the outer "convective" layer of the Sun, which is some 150,000 kilometres deep, it heats the gases within it, and causes them to rise, expand, dissipate the heat, and sink again. Observations show that above this layer is a "grainy" surface composed of individual cells of rising and falling gas roughly 500 kilometres deep. This is the visible surface of the Sun upon which dark blotches called sunspots may be seen from time to time. Sunspots are simply cooler regions of gas, which as a result of their temperature appear darker than the rest of the surface. The number of sunspots increases and decreases over a period of 11 years, marking a cycle of solar activity.

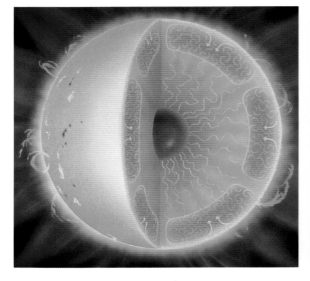

▲ **Solar cut-away** An artist's impression of the Sun, showing the core where energy is produced. The energy is transmitted up through the radiative zone into the outermost zone, where convection in the cells generates magnetic fields that show up as sunspots and prominences.

▶ **Invisible Sun** At X-ray wavelengths, the Sun emits invisible high-energy radiation. The spectacular false-colour image here was obtained in October 1991 by the Japanese Yohkoh solar observatory. The brightness and hue of the colour corresponds to the intensity of X-radiation.

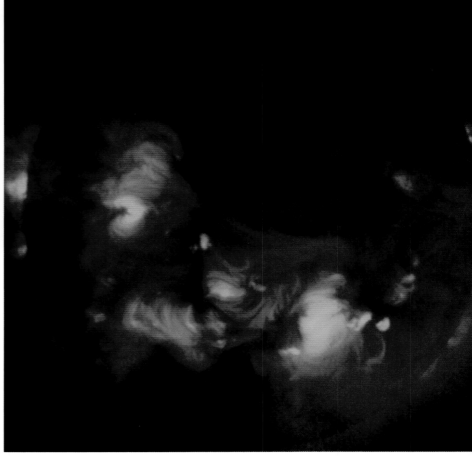

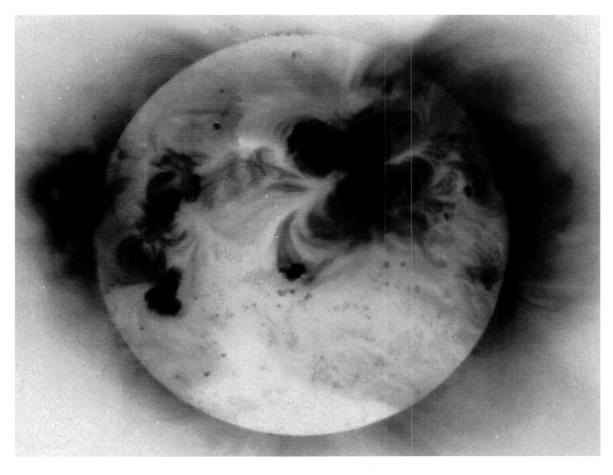

◀ **Corona** The Sun's outer atmosphere, the corona, can be seen in this image from the Soft X-ray Telescope aboard the Yohkoh solar observatory. X-rays are emitted from all parts of the corona, with brighter areas extending more than 1 million km above the surface. The features of the "invisible" Sun are made visible: large, irregular loops, bright points, and coronal holes.

▶ **Magnetic Sun** These two images show the Sun's visible surface (left) and a false-colour magnetogram displaying the magnetic polarity of the solar surface (right). Yellow areas are north poles, while blue and pink regions are south poles. Sunspots in the visible-light image clearly correspond to features of the magnetic structure. Sunspots are believed to result from local magnetic fields, which prevent heat rising to the surface.

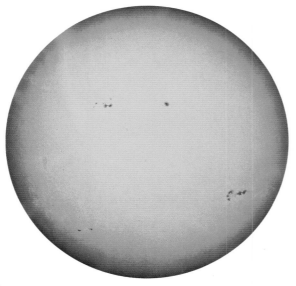

The Sun's photosphere, as we have said, effectively marks a boundary beyond which it is not possible to see. But one way to look at the interior is to monitor neutrinos – strange, almost ghostly subatomic particles – which are emitted deep inside the Sun's core during the nuclear fusion process. It is, however, very difficult to detect neutrinos and their unusual properties are poorly understood. So far, it has not been possible to obtain a picture of conditions within the core by observing them.

Scanning with sound

By far the most promising technique for probing the solar interior is known as helioseismology. Astronomers have known since the early 1960s that the surface of the Sun appears to oscillate by about 50 kilometres five or six times every half hour. These movements are caused primarily by acoustic compressions (like enormous sound waves) within the Sun, which combine to make its visible surface oscillate in regular patterns.

There are three basic methods of detecting solar oscillations. The first involves measuring the Sun's temperature very accurately; as the solar surface expands and contracts, its temperature changes, causing its brightness to alter by a few parts per million. A second method is to observe directly the minute changes in the Sun's diameter as oscillations occur. But by far the most common technique is the third: measuring the speed of the Sun's surface as it moves to and fro. This is made possible by the Doppler effect, which lengthens the wavelength of solar spectral lines as the surface contracts, and shortens them when it expands. The effect is much like the change in the tone of a passing locomotive, which sounds higher-pitched as the locomotive approaches and lower-pitched as it recedes. A little mathematical trickery allows the observed change in wavelength to be converted into measurements of the speed of the surface.

These oscillations of the solar surface are not just a localized effect; they are caused by sound waves generated inside the Sun by the convective motions of ascending and descending gases. Sound waves literally bounce around the solar interior and are reflected off the interior of the photosphere. The patterns they make on the surface are "standing waves", like the vibrations on a drum skin seen in slow motion. In much the same way that geologists learn about the Earth's interior by monitoring earthquakes, so solar scientists are able to probe the mysteries of the solar interior by observing these oscillating patterns on the Sun's surface.

The strange behaviour of the Sun does not end with oscillations. One of the greatest puzzles in current solar science is the paradoxical fact that the Sun's outer atmosphere is hotter than its surface. The exact extent of the Sun's atmosphere varies, but it can be as much as 15 times the Sun's visible diameter. From Earth, the solar atmosphere can be seen well only during total eclipses — when the Moon passes in front of the Sun. Unfortunately, these are rare events. In the 1930s a technique was invented that mimics solar eclipses by using a metal disk to block out the Sun. Coronagraphy, as it is known, was first used from high-altitude sites, where light scattering from dust in the Earth's atmosphere did not interfere too much with the observations; but with satellites, coronagraphy has been producing really useful results.

Coronagraphy and eclipse observations have revealed that the Sun's atmosphere is made up of several distinct zones. In the 2,000 kilometres nearest the Sun's surface, the temperature of the atmosphere rises to around 8,000°C. This lowest layer (in which the density of the gas thins considerably) is known as the chromosphere, but is usually not

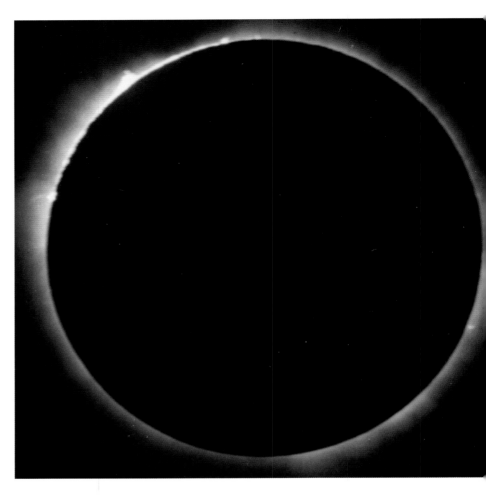

visible because the photosphere is a million times brighter. Just before the "total" phase of a solar eclipse, the chromosphere appears as a thin red arc around the Moon's edge. Spectroscopy has shown that this colour is caused by strong hydrogen emission — demonstrating the presence of hot hydrogen.

A startling rise in temperature occurs in the "transition region" in the top 500 kilometres of the chromosphere. Temperatures here rise from 8,000°C

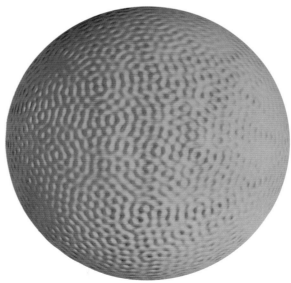

▲ **Total eclipse** *The Sun's chromosphere lives up to its literal meaning, "globe of colour", in this image of a solar eclipse recorded by astronauts aboard the Skylab space station in 1973. A distinct pink-red colour is evident, and small solar prominences are visible at top left.*

◀ **Solar golf ball** *A computer simulation of the oscillations of the solar surface. The surface wobbles in a complex manner as a result of sound waves bouncing around within the Sun. Observation of these movements gives solar scientists important information on the Sun's structure. The height of the oscillations has been exaggerated 1,000 fold in producing this image.*

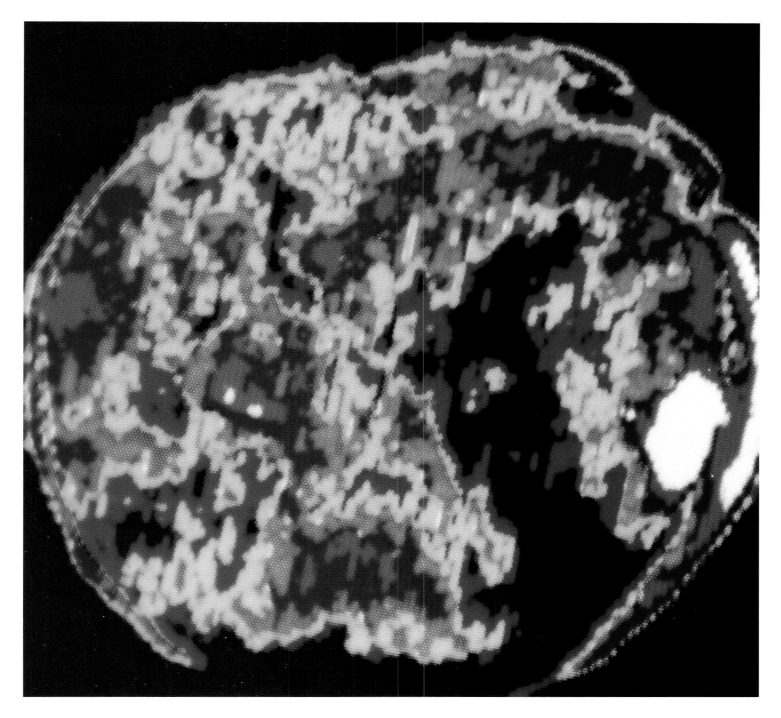

▲ Coronal hole *A false-colour image of the Sun at extreme ultraviolet wavelengths revealing a large coronal hole, which appears as a dark band on the right. Caused by the thinning of the Sun's constantly changing magnetic field, coronal holes allow solar wind particles to leave the Sun's surface. The image here was taken from Skylab in 1973.*

to 500,000°C, and astronomers are at a loss to explain why. The region marks the beginning of the extended outer atmosphere of the Sun, known as the corona, which has a very low density (roughly one atom per cubic metre) and a temperature of between 1 million and 2 million °C.

The corona extends far into space, and during an eclipse it is seen as a ghostly-white halo around the Sun. It changes its appearance during the 11-year cycle of solar activity: during peaks in the solar cycle, it appears spiky and jagged, with distinct plumes of activity; in quieter periods .it is more even and rounded. Closer inspection reveals that this jaggedness is caused by luminous gas erupting

out of the chromosphere. These solar eruptions are known as prominences. When viewed at the edge of the Sun they appear as loops and arcs – caused by electrically charged material following the curved lines of Sun's magnetic field. When seen on the disk of the Sun, prominences appear as dark silhouettes against the brighter solar surface.

The solar wind

Observations from space sometimes reveal large holes in the corona, whose presence is intimately linked with changes in the Sun's magnetic field. Normally these coronal holes are found close to the poles, but they can extend as far as the solar equa-

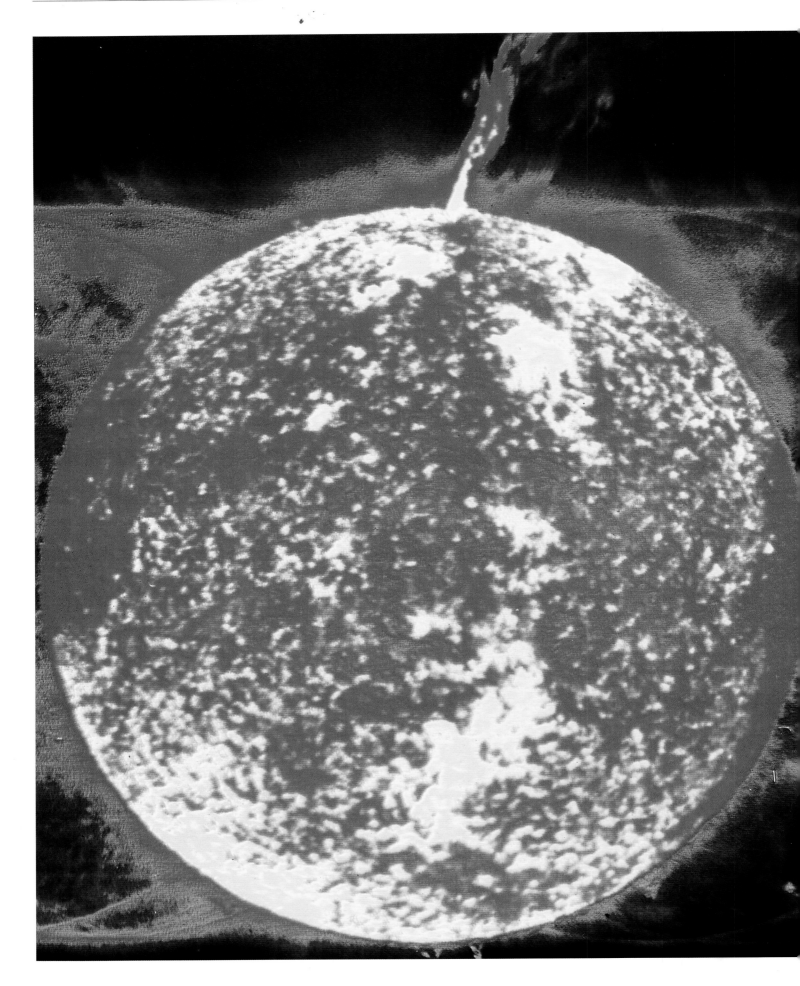

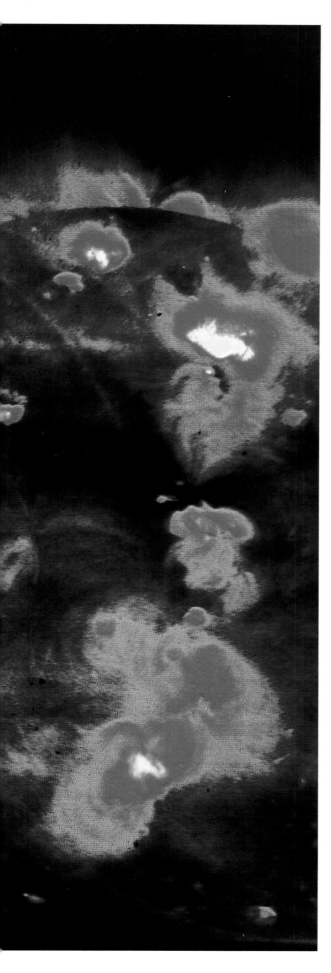

tor. Most importantly they seem to act as a source for streams of highly energetic particles, such as electrons and protons, as well as heavier ions and atomic nuclei ripped apart by heating of the Sun's outer atmosphere. This cocktail of highly energized, electrically conducting particles (technically known as a plasma) is called the solar wind.

The term solar wind is a slight misnomer, for it is quite unlike the winds found in the Earth's atmosphere, which blow because of differences in atmospheric pressure. In the vacuum of space, the solar wind consists of between one and ten particles in each cubic centimetre – hardly enough to create much pressure difference. The solar wind blows because the particles in it are highly energized and so move very fast; also there is nothing in the vacuum of space to slow them down. The solar wind extends throughout the Solar System, and spacecraft out beyond Pluto have detected it influence, which may stretch out as far as 150 times the Earth's distance from the Sun.

Because the corona is irregular, so too is the solar wind, and at times it can become quite gusty. Particles in the solar wind travel at speeds between 300 and 1,000 kilometres per second, with the faster particles being generated by particularly energetic bursts of solar activity.

The direction of the solar wind is under the control of the Sun's magnetic field, which rotates along with the Sun. This rotation means that the solar wind is ejected in the same way that water is from a rotary garden sprinkler. The result is a complex hotchpotch of twisting and spiralling magnetic field lines within which the solar wind is trapped.

Until recently, our view of the solar wind and magnetic field have been affected by our position in space. To get a picture of the whole of the Sun from the Earth is rather like trying to predict the Earth's weather by flying round the equator; observations are limited to within 10 degrees of the plane in which the Earth orbits the Sun (the ecliptic). There was an obvious need for a spacecraft to observe the Sun from outside the ecliptic, high above the Sun's poles, where, it was hoped, the structure of the solar magnetic field would be much simpler. Some theorists believe that the Sun's magnetic field-lines at the solar poles act as funnels for particles from elsewhere, such as the highly energetic (and still enigmatic) cosmic rays.

◄ **Solar prominence**
A vast prominence, billowing from the Sun across many millions of kilometres of space, captured in 1973 by Skylab astronauts using an ultraviolet camera. Colours have been added to the originally black-and-white image in order to bring out important detail and differences in brightness.

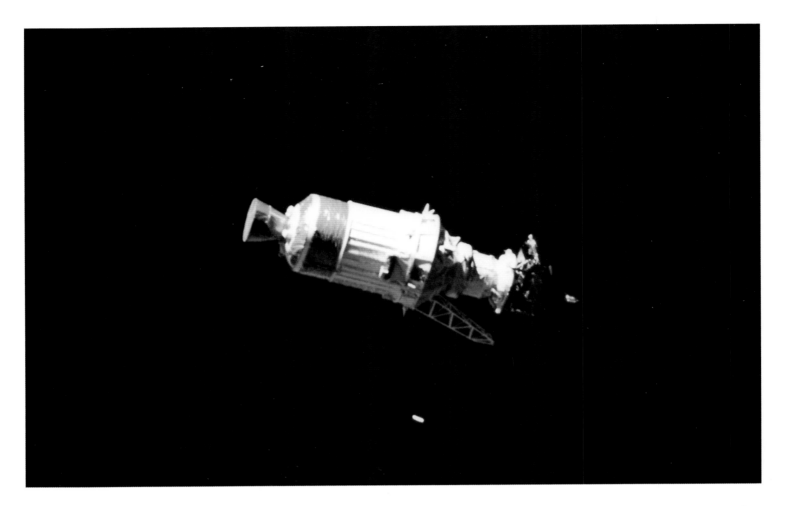

▲ **On its way** The Ulysses spacecraft was launched on 6 October 1990 from the payload bay of the space shuttle Discovery. At first scheduled for 1985, a series of delays pushed the launch back into the next decade. The powerful Inertial Upper Stage booster (left side of image) took Ulysses from Earth orbit to Jupiter in only 16 months, from where it harnessed the giant planet's gravity to slingshot it back towards the Sun.

The odyssey of Ulysses

A space mission is now well under way to add the missing third dimension to our studies of the Sun. It was launched in October 1990 from the space shuttle *Discovery* and is called Ulysses — the Roman name for Homer's mythological explorer Odysseus. The Ulysses space probe was built by the European Space Agency (ESA) and is the mortal remains of a larger project involving two spacecraft, the second of which was to have been built by NASA. Unfortunately, financial cutbacks led to the cancellation of a large part of the US contribution to the programme, and the *Challenger* accident delayed the start of what was left of the mission.

Ulysses' journey to the Sun's poles, and out of the plane of the planets, has meant a long and circuitous voyage. To push it out of the plane of the ecliptic, Ulysses had to make use of the gravitational field of the largest planet in the Solar System — Jupiter. Engineers at the Jet Propulsion Laboratory have long since refined the technique known as the gravity assist, in which a spacecraft "steals" gravitational energy from a planet using a slingshot technique. The "stolen" energy gives the spacecraft the necessary speed to go on to its next target. Instead of having to carry prohibitively heavy (and expensive) rocket stages to boost them towards distant

targets, spacecraft are now routinely accelerated in this way, invariably at the cost of a more circuitous and time-consuming route.

In February 1992, Ulysses reached Jupiter and passed within 450,000 kilometres of the Jovian north pole in such a way that it was accelerated rapidly downwards under the Jovian south pole and was kicked into a polar orbit of the Sun. As a result of this slingshot process, Ulysses became the fastest probe in space. From late June to early November 1994 Ulysses passed over the south pole of the Sun; it then travelled on to pass over the north pole in June to November 1995.

Initial results from Ulysses have confirmed some of the theorists' suspicions. A persistent coronal hole has been found over the Sun's south pole. Here, with no magnetic field loops to impede its progress, a high-velocity solar wind streams out of the Sun. Also, a far better understanding has been obtained of the expulsions of mass from the Sun, which are linked to the auroras and magnetic storms on Earth. There is more information to come and, providing there are no problems, Ulysses should still be operating in 2000 and 2001 when two further solar fly-bys are scheduled. These should be particularly interesting, because they coincide with the next maximum in the sunspot cycle.

▶ **SOHO** An artist's impression of ESA's Solar and Heliospheric Observatory (SOHO, scheduled for launch in late 1995 by an Atlas rocket from Cape Canaveral. It will orbit at a Lagrangian Point – where the gravitational attractions of the Sun and the Earth balance exactly – allowing the probe to remain stationary relative to both bodies. It will monitor the Sun, providing information on the behaviour of the solar atmosphere.

HOME PLANET

It is customary to portray planet Earth as a world unique in the Solar System. Among other things, it has a relatively dense atmosphere, which largely shields us from meteorites, and it appears to be the only planet on which continents drift on "tectonic plates". But above all else, ours is a water- and oxygen-rich world that is the only oasis of life in our planetary system. Yet the laws of physics that pervade the Universe and are obeyed on the other planets of the Solar System apply on Earth, and the examples above are merely manifestations of them. Three decades of planetary exploration have enabled us to place the Earth in its true physical perspective

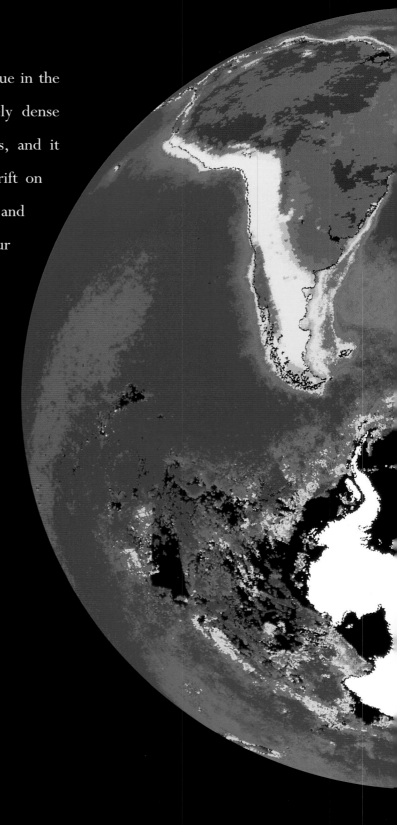

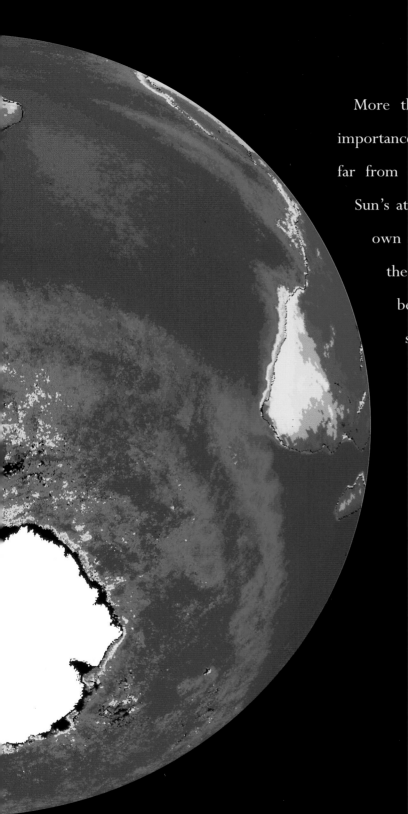

More than anything else, perhaps, we have realized the importance of our planet's interaction with the Sun. Space is far from empty, for the extensive outermost layers of the Sun's atmosphere have an influence that permeates into our own atmosphere. We are protected from the solar wind, the constant stream of particles from the Sun, only because we live under the electrically insulated atmosphere of the Earth. It is interesting to reflect that this atmosphere, while protecting us from the Sun, largely isolates us from the rest of the Universe – a Universe that is, for the most part, electrically conducting and filled with particles and radiation.

Fragile world The 1990s are a remarkable decade for Earth sciences, with a new generation of satellites dedicated to monitoring our environment. The changing extent of the polar ice caps (the southern ice cap is seen here) and the ozone layer are among the many problems that are in urgent need of investigation.

◀ **Southern Lights**
In April 1994, astronauts aboard the space shuttle Endeavour *took this photograph of the aurora australis or "southern lights". From low Earth orbit, this display of the solar wind's interaction with the Earth's magnetic field has the appearance of a vast luminous curtain of light. The predominant green-blue colour is caused by emission from energized oxygen atoms. Auroral displays normally occur at altitudes of about 100 km.*

The solar wind compresses Earth's magnetic field on the sunward side, and on the night side it stretches it into a long tail, in much the same way that the motion of a boat through the water causes a wake. The sunward boundary of Earth's field is known as the bow shock, and marks the limit of the Earth's magnetic influence. It appears where the solar wind particles pass from supersonic to subsonic speeds. This normally happens within a few Earth diameters of the Earth's surface, although the distance varies with the "gustiness" of the solar wind. Within the confines of the bow shock, where the Earth's magnetic influence is dominant, is the vast cavity known as the "magnetosphere".

Magnetic energy is stored in the magnetosphere, and can be released very dramatically, for Earth's magnetosphere and the solar wind do not interact smoothly. The electrical power created results in auroral displays and generates the Van Allen Belts, which trap radiation from the Sun.

The value of making observations in space was demonstrated by the discovery of these belts by the first successful US satellite, Explorer 1. Later missions revealed that the Van Allen Belts act like engines, pumping particles around the Earth and accelerating them. Particles may be trapped within the belts for anything from a day to 20 years, and are often released only in the surge of downwind activity that results from solar storms.

The deceleration of the solar wind resulting from the bow shock heats the solar-wind particles and makes their flow turbulent. Most of the particles are deflected past the Earth within the magneto-sheath (the region beyond the outer edge of the magnetosphere), but some seep directly into the magnetosphere and are funnelled down over the poles, producing spectacular auroral displays.

Inside and out

The Earth has the strongest magnetic field of all the inner planets, and it also has the most active geology. The reason for both is the Earth's hot inner core, a sphere of iron, which is probably hotter than the surface of the Sun but is kept solid by the intense pressure at the centre of the Earth. The solid iron core is surrounded by a layer of molten iron, and it is this molten layer that is the source of the Earth's magnetic field, for its internal motions make it behave rather like an enormous liquid dynamo. Our planet is rich in radioactive elements such as uranium

▼ **Cluster around** *At the end of 1995, the European Space Agency plans to launch four identical space probes called Cluster, which are designed to explore in detail the Earth's magnetic environment. They will fly together in formation, as shown in the artist's impression below. Their relative movements will enable scientists to make examinations of subtleties in the structure of the solar wind and the way that it interacts with the Earth's magnetic field.*

and thorium, and their decay is the main mechanism by which the core is heated. It is the flow of this heat outwards to the mantle and crust that drives many of the Earth's geological processes.

Although the Earth's surface has been sculpted by processes similar to those that shaped the other inner planets, the Earth is the most geologically active and has been continually altered from within. Most surface rocks are relatively young, dating from within the last 1,000 million years. The comparatively thin crust has evolved into "tectonic plates", which float on the mantle. The idea of plate tectonics, as the movements of these plates are known, was not widely accepted until the 1960s.

Volcanic features have been found on all the inner planets. On Earth, they tend to be found in regions overlying "hotspots", or where continental plates are pulling apart or coming together.

Volcanic explosions, are easily seen from space. In June 1991, for example, the Philippine volcano Mount Pinatubo erupted and vented an estimated 20 cubic kilometres of sulphur-rich dust and ash high into the stratosphere. Infrared sensors on board weather satellites detected this plume of material circulating around the Earth from its telltale thermal emission.

The eruption of Pinatubo has served to complicate any emerging consensus on the vexed question of global warming because its ash cloud remained suspended in the upper atmosphere for the rest of 1991 and during 1992. By screening the rays of the Sun, Pinatubo's ash cloud appears to have reduced the average temperature of the Earth's surface by about 0.5°C. This sort of random "background noise" makes it difficult to detect any underlying trend of global warming.

◄▶ Mount Pinatubo
The major eruption of the
Philippine volcano in June
1991 claimed 350 lives.
The remarkable image here
shows that one farmer was
oblivious to it all. The
eruption threw into the
atmosphere an estimated
20 cubic km of sulphur-
rich ash, which spread
around the world.

**▶ Volcanoes, Hawaiian-
style** In October 1992,
astronauts aboard the
space shuttle Columbia
returned this view of the
large island of Hawaii and
its three main volcanic
domes. In the foreground is
Mauna Loa, on the far side
is Mauna Kea and, to the
right, Kilauea from which a
small puff of steam is just
visible through clouds.
Volcanoes of this "shield"
type – with a flat top and
shallow sides – have been
found on Mars and Venus.

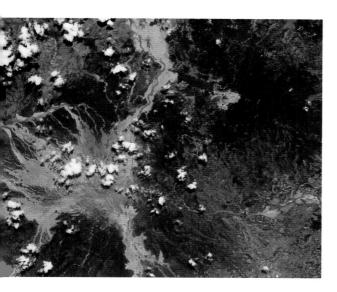

Satellite-borne infrared spectrometers reveal strong thermal emission that is characteristic of the presence of carbon dioxide, the most important greenhouse gas on Earth. A cocktail of other, less abundant, greenhouse gases is present, and calculations show that their collective effect keeps the Earth as a whole more than 30°C warmer than if they were not present. It is this that makes much of the Earth such an ideal habitat for life, but too much further warming would be detrimental. To determine past episodes in the Earth's climate, it is possible to drill into the polar ice caps and extract samples of ice containing tiny pockets of air trapped in past eras. To see how the concentrations of the greenhouse gases are changing now, however, requires systematic long-term studies from space.

What distinguishes Earth's atmosphere from that of other planets becomes apparent after taking temperature profiles from space. All planetary atmospheres cool with increasing altitude, except the Earth's, which warms considerably at an average altitude of 20 kilometres.

The reason for this is that there is a diffuse, yet detectable, layer of a highly unstable form of oxygen called ozone. (Ozone consists of three oxygen atoms combined to form a single molecule, whereas oxygen molecules in the atmosphere normally have only two atoms. Ozone forms essentially through the action of sunlight on oxygen molecules and is very unstable.) Ozone absorbs ultraviolet radiation very strongly, and thus the ozone layer protects us from this radiation, which is extremely dangerous – particularly in large doses – to living creatures on the surface of the Earth.

Thanks to humanity's polluting influence, the ozone layer is progressively being stripped away, particularly by a group of compounds known as CFCs (chlorofluorocarbons). The development each spring of a hole – a severe thinning of the layer –

over the south pole is very extensive. The latest generation of infrared spectroscopes in Earth orbit have revealed additional subtle, but disquieting, details about damage to the ozone layer, particularly over the northern hemisphere.

The ever-changing cloudscape of the Earth shows that, of all the terrestrial planets, ours has the most dynamic weather systems. The basic cause of atmospheric circulation is that air is warmed at the equator, then rises and travels towards the poles, where it cools and descends. The air does not reach the poles directly, for there are three "cells" of air between the equator and the poles, which complicate the airflows. Because of the Earth's rotation, these cells give rise to the trade winds, which drive much of the weather at mid-latitudes. Weather patterns seen on Mars and Venus obey the same basic laws of atmospheric science, although the phenomena show themselves in rather different ways and are, for the most part, less pronounced.

Sea changes

The fact that nearly three-quarters of the Earth's surface is covered by water has prompted the writer Arthur C. Clarke to question why our planet is called Earth, when it could more aptly be called Ocean. The oceans play a fundamental role in shaping the climate of our planet; for example, there is more energy stored in the upper few metres of the seas than in the atmosphere above. In effect, the oceans act as a global thermostat, although the manner in which they redistribute heat is imprecisely unknown. If we are to have any hope of predicting changes in global climate, we will need to understand the role of the oceans.

Oceanographers can draw only a limited picture using measurements from the surface of the seas. The best data come from the sea lanes along which ships regularly navigate. Although it is possible to take ocean research vessels almost any place there is enough water to sail them, a true global perspective of the oceans is made practical only from satellites, and these have revolutionized oceanography.

The aim for the 1990s is to get a "snapshot" of the state of the oceans against which future changes can be measured. A host of satellites are now in operation using microwave sensors to scrutinize the Earth's surface. Using microwaves allows the sensors to work at night and see through cloud.

In 1991 the European Space Agency (ESA) launched its first Earth remote-sensing satellite, ERS-1; this was followed a year later by the launch from French Guiana of a dedicated oceanographic satellite, a joint French–US project, known as TOPEX/Poseidon. Both satellites use radar altimeters which send pulses of microwave radiation

down to the Earth's surface; the time taken for the reflected signal to return gives the height of the ocean below to an accuracy of a few centimetres. Such measurements have shown that there really is no such thing as sea level, for the surface level can often vary by many metres across large basins.

TOPEX/Poseidon and ERS-1 cross-check and calibrate each other's measurements. Their different orbits determine the observations that each can make. ERS-1 orbits the Earth in a near-polar orbit, which means it can see the whole of the Earth's surface as the planet turns below. Its orbit is also Sunsynchronous, meaning that it passes over the same part of the Earth's surface at the same time every

day, making changes easy to monitor. The orbit of TOPEX/Poseidon is inclined at 66 degrees to the equator, but unlike ERS-1, it crosses the same part of the ocean at different local times. This allows it to make measurements in different tidal conditions, so that the underlying currents can be seen more clearly once daily tidal variations are removed.

Climate anomalies

With its ability to sample greater ranges of the ocean surface, TOPEX/Poseidon has received greater attention. It has given oceanographers new insights into the mysterious workings of a climate anomaly, the effects of which link Chilean fishermen, floods

▲ **Lighting and lightning** *Both artificial and natural lighting are visible in this photograph centred on Central America, taken by shuttle astronauts in May 1993. The cluster of lights in the middle is associated with Mexico City.*

▲ **Atmosphere from space** *The scattering of light by particles makes the atmosphere appear from space as a brightly lit ribbon of reds and blues. The dark region in the stratosphere on the left is probably volcanic ash from the Mount Pinatubo eruption.*

▶ **The hole in the sky** *Each southern spring, two-thirds of the ozone layer above Antarctica disappears for nearly six weeks. This computer-enhanced image shows ozone as it would appear if it absorbed visible and not ultraviolet light. The picture was generated from observations returned by the Nimbus 7 Total Ozone Mapping Spectrometer in October 1990.*

in California and forest fires in Australia. This phenomenon is the irregular current called El Niño. One sign of the onset of El Niño is a change in sea level, which can easily be detected by the altimeters aboard TOPEX/Poseidon.

El Niño has been seen to alter atmospheric circulation patterns on a global scale, resulting in flooding and drought in distant parts of the world. The cause seems to be relatively high-level heating of surface water in the central equatorial Pacific, where the oceans are already among the warmest in the world. The heating generates a warm current which can extend as far as the coast of South America. Because this usually occurs at Christmas time off the coast of Chile, local fishermen have

associated the phenomenon with "the boy child" of the nativity, El Niño. The events are irregular both in occurrence and in duration; the most recent started in the autumn of 1991, reached a peak in early 1992, and then progressively diminished.

Until recently, few observations of the sea surface in the central Pacific had been made, and it had been difficult to determine how the "source" of El Niño was evolving. In December 1992, TOPEX/Poseidon's observations of the region revealed the unexpected rise of a large "pulse" of water, known as a Kelvin wave, which effectively replenished El Niño. When these observations were incorporated into a computer model of the climate, operated by the US Naval Research Laboratory, the model predicted warmer and drier summers across the southern hemisphere. This resurgence of El Niño took many oceanographers by surprise, and led to the need for finer tuning of models. Only when these models can reproduce the patterns of the recent event will it be possible to predict future occurrences of El Niño with accuracy.

The temperature of the sea surface is the single most important factor in the Earth's climate, and accurate readings of sea temperature are essential for the computer models that predict changes in our weather. Accordingly, the ERS-1 satellite was equipped with a number of instruments dedicated to the task. The Along-Track Scanning Radiometer (ATSR) measures the infrared radiation emitted from the sea and converts the data directly into temperature readings. The ATSR has two fields of view: directly below and 57 degrees ahead. Because the satellite is moving, the same spot on the sea surface is scanned twice in rapid succession. In the few seconds separating each reading, the sea surface temperature will not have changed, and so if the ATSR records any temperature difference between the two readings it must result from atmospheric conditions. Thus, for the first time, researchers are able to discriminate between the effects of sea fogs and clouds, which formerly led to systematic errors in measuring sea temperatures.

The extent of ATSR data has revealed many previously unsuspected ocean phenomena, including enormous thermal waves that stretch right across the Pacific Ocean, the exact cause of which is unknown. El Niño is merely the most famous of a huge variety of such phenomena.

ERS-1 has also pioneered another technique, Synthetic Aperture Radar (SAR), which, as we shall see in a later chapter, has also revolutionized our picture of the planet Venus. SAR works by imitating the performance of a radar receiver much larger in size than the spacecraft. The craft beams down a wide swath of radar signals to the surface of the planet in question and picks up the echo a few seconds later. By this time, however, the spacecraft has moved along in its orbit, so the radar echo is picked up several kilometres away from where the beam was originally emitted. Thus the motion of the spacecraft synthesizes the resolution of a radar many kilometres in length. The radar echoes are then sorted by frequency and time to reconstruct a remarkably highly detailed radar picture.

SAR has allowed such phenomena as "internal waves" in the Earth's oceans to be seen from space.

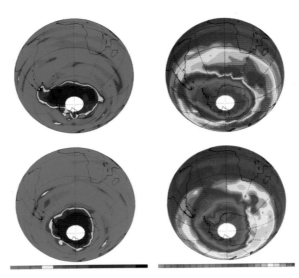
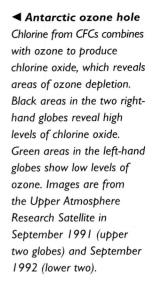

◄ Antarctic ozone hole Chlorine from CFCs combines with ozone to produce chlorine oxide, which reveals areas of ozone depletion. Black areas in the two right-hand globes reveal high levels of chlorine oxide. Green areas in the left-hand globes show low levels of ozone. Images are from the Upper Atmosphere Research Satellite in September 1991 (upper two globes) and September 1992 (lower two).

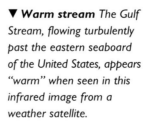

► Thunderball A 50-km-wide thunderstorm off the coast of Nigeria snapped in astonishing detail by shuttle astronauts in May 1993. Two major upwellings appear as cauliflower-shaped thunderheads at the heart of the storm.

▼ Warm stream The Gulf Stream, flowing turbulently past the eastern seaboard of the United States, appears "warm" when seen in this infrared image from a weather satellite.

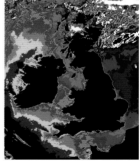

▲ Where to swim Great Britain's climate is heavily influenced by the warm Gulf Stream. The Advanced Very High Resolution Radiometer on the NOAA Tiros N satellite shows the Atlantic in April is a warm 13°C, while the North Sea is much cooler at 8°C.

▲ Caribbean blue
Shallow tropical waters off the coast of the Bahamas are seen in this view taken by shuttle astronauts. However, visual imagery plays only a small part in understanding the state and condition of the oceans.

These waves result from the mixing of currents of different saltiness – most notably in the Straits of Gibraltar. SAR images do more than provide fascinating images. The way the surface scatters the radar signal also reveals useful information. If you drop a tennis ball onto a completely flat surface, the ball will bounce straight back up; but if the surface is rough, it will generally bounce back at an angle. In the same way, the reflection of the radar beam from the SAR on board ERS-1 depends on the texture of the surface, and thus the disturbances due to winds and waves are readily visible from ERS-1.

The SAR on ERS-1 has been used to observe directly how the wind generates waves on the sea surface, an important consideration in weather forecasting. Until now, meteorological data from the oceans – particularly in the southern hemisphere – have been sparse as they have been gathered only from buoys and shipping lanes. But from its orbital vantage point, ERS-1 uses the SAR to look at the state of the ocean waves in what is known as its scatterometer mode. Radar signals are sent onto the ocean surface in three different directions, yielding a reflected signal that enables meteorologists to measure the state of the sea surface.

Previously, weather forecasters had to guess what was happening in areas where no measurements were being made. Now, ERS-1 data can be fed directly into global weather-forecasting models around the world. It is hoped that in the future a network of satellites will return the same kind of real-time data round-the-clock, and oceanographers plan to use the data to provide an up-to-the-minute view of the state of the oceans, known as the Global Ocean Observing System (GOOS).

Scientists also plan to use ERS-1 observations to measure the exact shape of the Earth – known as the geoid – with far greater precision than is currently possible. By monitoring the height of the Earth's surface over a period of time, it will also be possible to see how the Earth's tectonic plates are moving – information that will be of great value in predicting earthquakes and volcanic eruptions.

Polar observations

Until recently, glaciologists often complained that knowledge of the Earth's ice caps was less advanced than knowledge of the Moon's surface. This has now changed, thanks to ERS-1 and a new generation of polar orbiting satellites. One recent success of the SAR technique has been the discrimination of sea ice, packed snow and rock outcrops in the Arctic. This detailed information has led to a greater recognition of the importance of the polar regions in the patterns of Earth's climate. Of all the other planets in the Solar System, only Mars has similar polar ice caps, and their seasonal growth and shrinkage is known to have a fundamental effect on the way the Martian climate develops each year.

The polar ice caps effectively keep the Earth cool. Although less than 10 per cent of the Earth's surface is covered by ice, its presence limits the exchange of heat between the atmosphere and the ocean in polar regions. There is a fear that, if the Earth's climate is warming up, the melting of ice will serve to accelerate global warming, for there will be less ice

▼ **Wave heights** *From its near-polar orbit around the Earth, the European ERS-1 satellite makes repeated "strips" of radar observations of the sea surface. Shown here is the basic radar information that reveals how the wind generates waves across the oceans.*

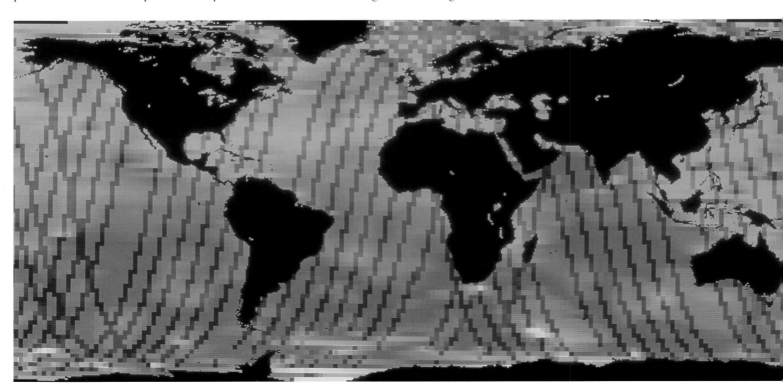

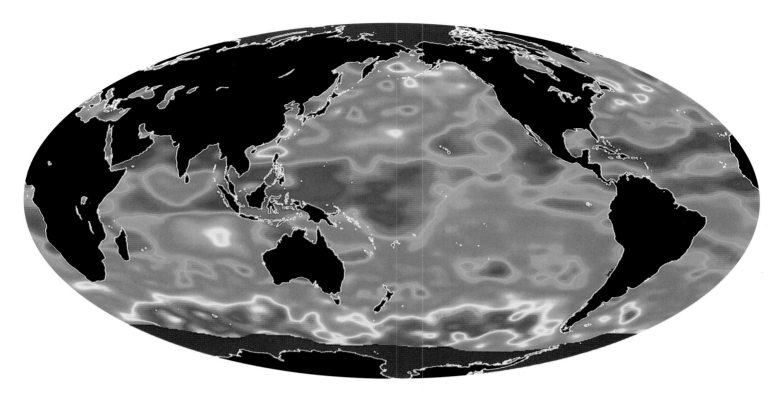

▲ **Wind speeds** In October 1992, the US TOPEX/Poseidon radar altimeter returned detailed information on wind speeds across the ocean surface. At radar wavelengths, calm seas appear "brighter" because they return a stronger reflection (they are shown here coloured blue). The strongest winds, of 55 km per hour (red colour), are found over the southern ocean. The trade winds are also visible in the subtropical regions.

remaining to reflect radiation back into space. The rate of heat absorption would then quicken over the Earth's surface, in turn causing more ice to melt, and leading eventually to a marked rise in the average sea level. Many low-lying countries, such as Bangladesh and the Maldives, would suffer almost untold economic and social damage.

Polar ice acts as a valuable marker for global climatic change: any changes in climate will be mirrored by conditions at the poles. Strange things can occur. In Antarctica, for example, parts of the ice sheets sometimes break off and float into the surrounding oceans, and seriously disrupt what is known as the "mass balance" of Antarctica – the annual gain and loss of ice due to seasonal freezing and thawing. Normally, the mass balance is equivalent to about 2.1 million million tonnes of water. In the autumn of 1991, a vast chunk of ice about the size of London, broke away from Antarctica. The giant iceberg (which was christened, perhaps not

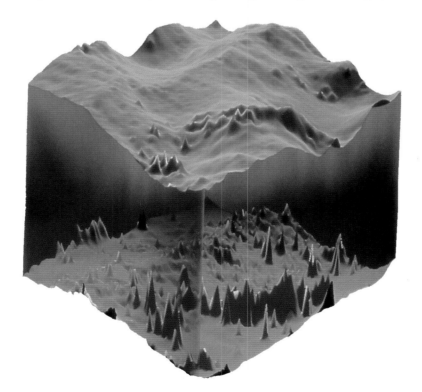

◀ **Ocean floor** Measurements of the "height" of the ocean surface (the top half of the picture) over many months reveal the underlying shape of the ocean floor (bottom half). In this image based on ERS-1 data, the vertical scale has been exaggerated many hundreds of times to show how ocean-floor ridges affect the flow of surface currents.

surprisingly, The Mother of all Icebergs) contained some 2.9 million million tonnes of water, and succeeded in complicating weather systems globally. It was photographed by astronauts aboard the Space Shuttle *Atlantis,* and was the biggest iceberg in the South Atlantic yet observed from space.

Because of Antarctica's remoteness, observations from satellites such as ERS-1 provide the only realistic way of determining how the mass balance may be changing. Half the rise in average sea-levels recorded in recent years could be due to changes taking place in Antarctica. Indeed, Scientists now believe that the Antarctic ice sheet is shrinking. But more research is needed for us to determine what this will mean for our climate in the future. With the launch of ERS-2 in April 1995, observations should continue until at least the end of the twentieth century and probably beyond.

Is there life on Earth?

Some of the images on these pages reveal the world from a curious perspective: one that an alien visitor might have when examining our planet for signs of life. The most interesting of such photographs were obtained by the Galileo spacecraft as it sped by the

◀ **The Mother of all Icebergs** In September 1991, the space shuttle Endeavour *passed over Iceberg A24, which was then close to the Falkland Islands in the South Atlantic. When this photograph was taken, the huge iceberg measured approximately 30 by 60 kilometres.*

▼ **The Andes** *A photomosaic made from 42 images of South America where the Chilean, Peruvian and Bolivian borders meet, taken by the Galileo craft on its second fly-by of the* Earth in December 1992. Visible and near-infrared filters were used, enabling variations in vegetation and soil to be made out.

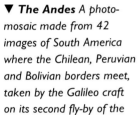

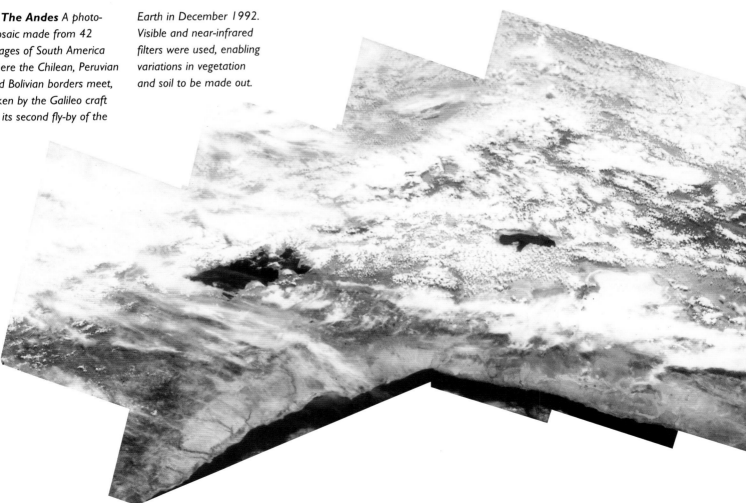

► **Farewell to Earth** *Antarctica is visible at the bottom of this view of Earth, which was taken from nearly 2 million km away, as dawn rises over the Pacific. The Galileo spacecraft shot the image on 22 December 1992, when the probe began the final leg of its journey to Jupiter, after cranking up its speed by "stealing" energy from the Earth.*

Earth in December 1990 on its way to Jupiter. While it was passing over Antarctica, the spacecraft's infrared spectrometer observed tell-tale signs of the photosynthesis (the process by which plants and algae produce energy from sunlight) caused by the massive presence of the marine organism phytoplankton. Over Australia, Galileo's television cameras searched in vain for the presence of regular, artificial surface features, such as roads or cities, but was not able to find them. Other instruments did find spectral signatures of gases such as methane and nitrous oxide, both typical products of biological activity. Observations of the relative abundance of oxygen also could not easily be accounted for by simple chemical means. Similar results were obtained when Galileo made a second pass in December 1992, although the track of the craft again did not pass over the more heavily populated northern hemisphere. As more than one Galileo scientist has said, observations from space revealed that life clearly existed on Earth, but it was difficult to tell if it was intelligent life.

► **Antarctica** *The ERS-1 satellite passes within 8 degrees of the South Pole, which it cannot image, so in this altimetric map the centre is missing. The relatively smooth, dome shape of the Antarctic icesheet is evident from this processed image.*

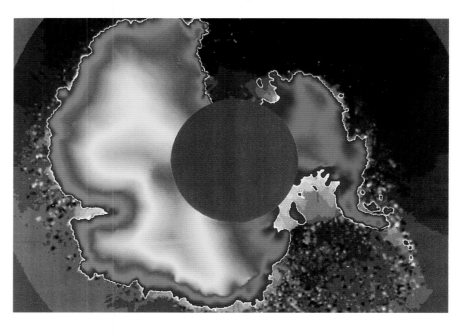

First Rung on the Ladder

In the spring of 1994, the Moon was in the news again after a lull of over two decades. Astronomers had long speculated that water ice might exist in the bottom of deep craters at the lunar poles, where the surface is in permanent shadow. Evidence for this ice was reported by the first American spacecraft to be launched to the Moon in 21 years. Known as Clementine; it was a small, automated probe built by the US military using hardware developed as part of the "Star Wars" Strategic Defence Initiative. If Clementine's observations are confirmed, the presence of water ice suggests that the Moon was bombarded by comets early in its history. Kept permanently in shadow within craters, the ice remained frozen for aeons as a pristine record of the Moon's early history. Some twenty-five years after the first humans walked on the Moon, it is clear that there is still a great deal more for us to learn about our nearest neighbour.

The Earth's natural satellite is on average only 385,000 kilometres from us, and keeps the same face turned to us at all times. It's diameter is just over a quarter of the Earth's, and it has about two per cent of the Earth's volume and little more than one per cent of its mass. This still makes the Moon far larger in proportion to its parent planet than all but one other moon in the Solar System – Pluto's moon, Charon.

The lunar surface is a mixture of heavily cratered highland terrain and darker, smoother lowlands called maria (Latin for "seas", which they were once thought to be). Most of the craters were produced in the first 1,000 million years of the Solar System's existence. Volcanic activity then covered parts of the surface with lava, filling in many of the ancient craters. For a period, the maria were indeed liquid seas of hot lava, but they eventually cooled and solidified, and were peppered by more recent, less frequent cometary impacts.

Past and present Over twenty-one years separate the time when astronauts of the Apollo 17 mission – the last to walk on the Moon – photographed the view (left) of the large crater Eratosthenes and the date of the remarkable image (right) of the Earth rising above the Moon, which was taken by the Clementine spacecraft as it orbited the Moon in 1974

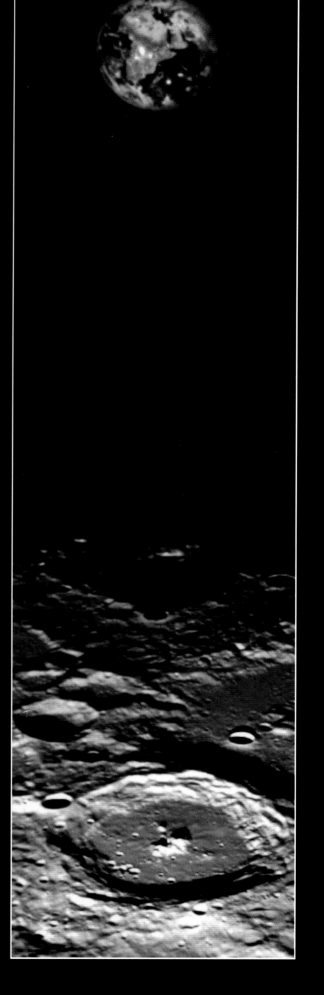

tated by a balance between the cooling of its interior and the energy generated by the impact of bodies and the radioactive decay of rocks. Within 100,000 years of the Moon's formation, the outermost lunar layers were molten, but they soon solidified as a crust to form the older, rugged terrain that is still visible in the lunar highlands. Within the first 500 million years, the Moon was bombarded from space; some objects were large enough to carve out giant impact basins, many hundreds of kilometres across. During this period, interior heating caused by radioactive decay was greater than the heat lost as the crust cooled and, where the crust had been weakened by impacts, molten material spewed out as lava and spread across the surface. The surface samples show that this lava was runny enough to spread very thinly for many hundreds of kilometres, unlike the more viscous lava flows seen on Earth.

Rock samples from the maria are definitely volcanic in origin. The dark, fine-grained material is similar to terrestrial basaltic rock. In other ways the lunar samples are quite different: they completely lack the water and other volatile materials that are often bound up chemically in Earth rock. Many of the basalts found on the Moon contain so many minerals that they would be classed as ores on Earth. The older, highland material is richer in aluminium and shows signs of being repeatedly pulverized and fused together in a molten state. Called "breccias", the basaltic rocks date from the earliest epochs of lunar history.

▲ **Moon rocks** In total, 382 kilograms of lunar rock was returned by the six Apollo craft that landed on the Moon. Only a small proportion has ever been examined, and the majority of it remains sealed in a nitrogen-rich depository at the Johnson Space Center.

◀ **First steps** When the lunar module Eagle landed in the Sea of Tranquillity, humanity accomplished what is perhaps its most remarkable achievement. The "magnificent desolation", as astronaut Buzz Aldrin described the lunar surface, was disturbed only by his and Neil Armstrong's bootprints and a few tokens of the visit.

The Moon after Apollo

Between July 1969 and December 1972, a dozen astronauts walked on the Moon. By analysing the rocks they gathered, geologists have been able to date when the surface-shaping processes mentioned above occurred. As a result, we now know when the bombardment suffered by the Moon, and many other Solar System bodies, took place. In all, 382 kilograms of lunar material was brought back by the Apollo astronauts. The Soviet Luna probes, the last of which, Luna 24, landed in August 1976, added a few hundred grams more. The Moon rocks showed, beyond all doubt, that the surface was very old and composed of a hotchpotch of lava flows and material pulverized by impacts. The ages of the rocks have been determined by isotopic analysis (a technique based on the rate at which radioactive atoms decay into other types of atom). The methods developed specifically for dating the Apollo samples are now widespread and are routinely applied to the meteorites that land on Earth from time to time.

It is now possible to assemble, with a reasonable degree of certainty, a chronology of how the Moon evolved. As with all planets, its evolution was dic-

▲ **End of an era** When the Moon-landing of the Apollo 17 mission ended in December 1972, astronauts Cernan and Schmitt left behind this plaque, to mark the conclusion of the Apollo space programme.

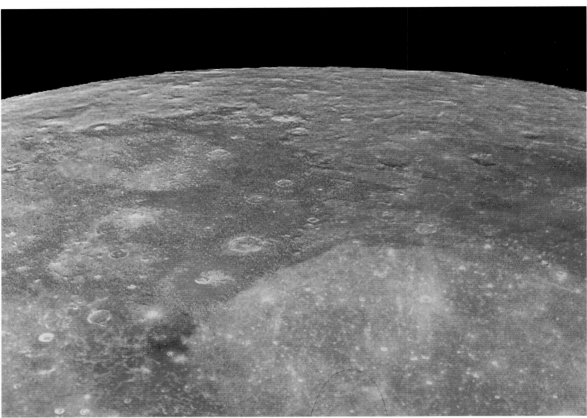

▼ **Sidelong glances** In 1990 and 1992, the Galileo spacecraft made two fast fly-bys of the Earth–Moon system. On the first of these, it took a sequence of images through violet, red and near-infrared filters to return an unusual sideways view of our celestial neighbour. The large feature left of centre is the Mare Orientale, which consists of concentric craters some 1,000 km across; the familiar maria of the lunar nearside appear on the right of the picture.

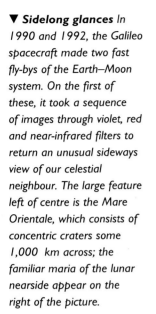

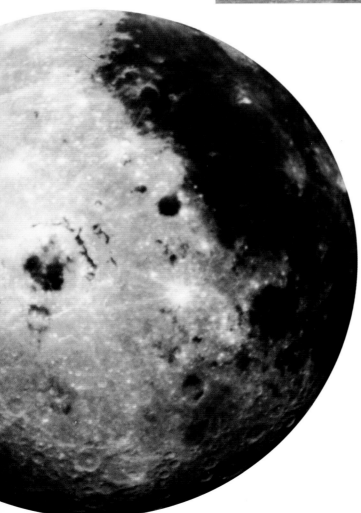

The Moon inside and out

But what of the origins of the Moon? The key was provided by chemical analysis of the Apollo samples. Before the Apollo landings, it was widely believed that the Moon split off from the Earth's outer layers, as it had a similar density to our planet's mantle. But the lack of volatile chemicals in lunar rocks alone precludes this and suggests that the Moon formed in a warmer region of the early Solar System, closer to the Sun, where such materials would have been "driven out". It is also highly unlikely that an object the size of the Moon could be captured into orbit by the Earth. No matter how slowly the Moon passed by, computer modelling of the gravitational interactions between the two bodies shows they would tend to disrupt each other's motion and fly off into new orbits around the Sun.

Thanks to the ever more sophisticated modelling possible with supercomputers, a theory has been developed that seems to resolve the main problem of the Moon's origin. Known as the Giant Impactor Theory or, less formally, as the "Big Splash", it suggests that an object the size of Mars scored a direct hit on the newly formed Earth. The impact melted the outermost layers of the Earth's crust so that the core of the impacting body disappeared into the core of the Earth. At the same time the shock of the impact splashed a molten ball of material into space; this ball eventually coagulated

▲ **The blue, red and yellow Moon** On its second fly-past of the Moon, Galileo's ability to observe in the near-infrared gave a foretaste of its encounters with the moons of Jupiter. The colours in this photograph have been chosen to bring out lunar features clearly. Dark orange and red areas represent the lunar highlands, while, paler orange colours indicate the titanium-poor basalt rocks of the Mare Serenitatis on the lower right. The Mare Tranquillitatis to the left appears mauve because of its relative richness in titanium. The purple region to the left of the centre marks deposits from past volcanic activity.

to form the Moon. This molten material would have been so hot that many of the volatiles within it boiled off. The "Big Splash" theory has its opponents, mainly because it requires a chance collision; but it leaves fewer problems unresolved than previous attempts to explain the Moon's origin.

Another related problem is how the core of the Moon has evolved over time. Today, the Moon has no detectable overall magnetic field, yet some of the rock samples returned to Earth were definitely formed in the presence of a strong magnetic field. Did the Moon once have an active core that has since solidified? The simplest way to probe inside the lunar interior is to make seismic measurements. By measuring how shock waves propagate through its interior, it is possible to come to a number of conclusions about the Moon's structure. Many moonquakes were recorded by the four seismic stations left behind by the Apollo astronauts, some occurring at shallower depths than others. The upper stage of the Lunar Module was routinely jettisoned just before the astronauts left lunar orbit for home, and seismic measurements of its impact on the surface provided additional information.

Molten to the core?

The Apollo seismic measurements, coupled with more recent investigations, have enabled geologists to piece together the following picture of the Moon's interior. The crust is thicker and more rigid than that covering the Earth, and it has not split into tectonic plates. On average the crust is about 70 kilometres thick and is generally thicker on the far side. Beneath the crust is a mantle, which probably extends down for another 800 kilometres, and it is in this layer that moonquakes seem to originate. In some areas, the mantle has bulged up to distort the local gravity field measured at the surface. These mass concentrations (mascons, for short) have been detected by their minute gravitational effect on the orbital motions of spacecraft above them.

Some scientists have suggested that the Moon has a hot, perhaps partially molten, core with a sizeable iron content; but none of the Apollo experiments support this idea. If there is a core, it cannot be very large. The relatively low density of the Moon and calculations of its moment of inertia (a measure of how smoothly it rotates) suggest that its interior is fairly uniform in composition. At most, the core may be 400 kilometres in extent, comprising at most only 4 per cent of the total lunar mass.

One further puzzle, which is probably related to this aspect of lunar evolution, is the Moon's asymmetry. Laser measurements from the Apollo spacecraft of the "height" of the Moon's surface show that the Moons is distinctly lopsided, with the far side

▲▼ **Blue Moon** Low-resolution mapping from the Clementine probe reveals subtle differences between the Moon at visible wavelengths (above) and in the ultraviolet (below). Clementine carried a series of sensors operating at wavelengths that had never before been used to study the Moon at close quarters.

bulging further outwards than the near side. For the most part, the far side is composed of light coloured, heavily cratered terrain, which is much older than the darker material found in the extensive maria on the nearside. The crust on the far side is thicker, and there has been less volcanic activity. Why this is so remains elusive. Unfortunately, we have no rock samples from the far side – all the Apollo landing sites were on the nearside, close to the edges of the maria for safety reasons and, because of fuel limitations, close to the lunar equator.

New views

The last humans to walk on the Moon were the crew of Apollo 17 in December 1972, and since then no astronauts have been back. In 1977 when the Apollo seismometers were switched off to save money, lunar exploration came to an abrupt and undignified halt. The Japanese launched their Hagomoro probe to the Moon in January 1990, but this was not entirely successful. It was only in December 1990 that lunar exploration underwent a brief revival when the Galileo spacecraft made the first of two fast fly-bys of the Moon, observing it from unusual angles. The first encounter extended spectral measurements and colour mapping across much of the far side. And by comparing the infrared and visible readings from one of its spectrometers, it was possible to get reliable estimates of the distribution of certain surface minerals. For the first time, Galileo clearly observed a large, depressed region known as the Aitken basin, a feature previously only suspected from Earth-based observations. Galileo found that material in this basin was darker than the surrounding area, and

presumably richer in iron and magnesium. The material could either have come from the underlying mantle, or perhaps from an ancient flood of lava that occurred earlier in lunar history than the lava outflows accompanying the formation of the maria.

In its second lunar fly-by two years later, Galileo passed over the lunar north pole and examined a number of craters that are permanently shadowed from sunlight. Some authorities believed that water ice could exist within these craters – material surviving from comet impacts early in the Moon's history. But because most of Galileo's instruments observe reflected sunlight, they could not see into the dark craters to clarify this matter. However, in 1994 tantalizing observations from a new probe, Clementine, gave hope to those who had predicted the presence of water.

Clementine arrived in lunar orbit in February 1994 and, in May, was due to head off to the asteroid Geographos. Unfortunately, because of a computer error most of Clementine's fuel was wasted and it will now never reach its goal. Before this mishap, however, Clementine did manage to return some useful observations of the Moon.

In March 1994 Clementine directed its radar at the Moon's north pole and on to those mysterious areas that are permanently in shadow. When observed by NASA's Deep Space Network, the reflected signal suggested the presence of ice. But the results were far from conclusive, and a NASA spokesman has recently pointed out that too much was being made of the idea of frost on the Moon. There will, however, doubtless be further surprises for the next generation of lunar orbiting spacecraft – or, for that matter, the next human lunar explorers.

◀ **Moonrise** In October 1992, astronauts aboard the space shuttle Columbia returned this stunning image of the newly risen Moon.

▲ **Lunar pole** The southern pole of the Moon, which was not seen at all by the Apollo missions, was mapped in detail for the first time by Clementine.

CAUGHT IN THE GLARE

When the first clear pictures of the planet Mercury were returned by the Mariner 10 probe in 1974, they revealed a barren, crater-scarred world very similar to the Moon. The pictures were hardly spectacular and caused many people to dismiss the innermost planet as an uninteresting place. Now, with the mass of information accumulated from the exploration of the rest of the Solar System, it is clear that Mercury may hold important clues to the violent birth of the other planets and of the Earth itself.

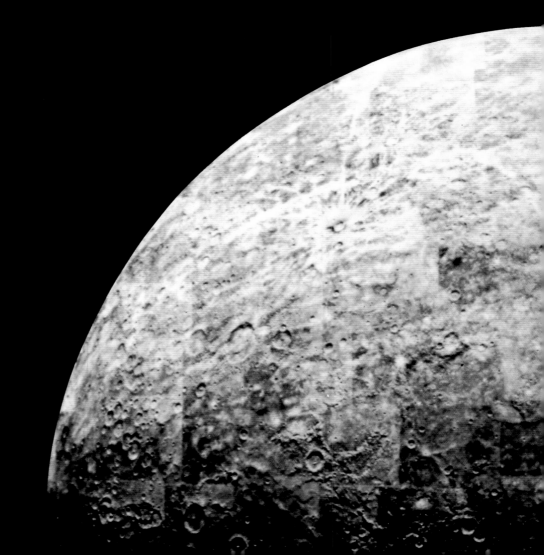

Mercury is a small world, with just over 5 per cent the volume and mass of the Earth and virtually no atmosphere. Before the Space Age, telescopic observers knew little more about the planet than the bare mechanics of its orbit. As Mercury is the closest planet to the Sun, it has the shortest year and takes only 88 days to complete one orbit. Yet this orbit is so elongated that, at its closest, Mercury comes to within 46 million kilometres of the Sun and, at its furthest, is some 70 million kilometres distant.

Pockmarked surface
The crater-strewn surface of Mercury was photographed for the first time by the Mariner 10 spacecraft in the mid-1970s. Pictures of the planet are "bleached" by the sheer intensity of sunlight, so in this shot a tint has been added to approximate the planet's visual appearance in slightly less glaring conditions.

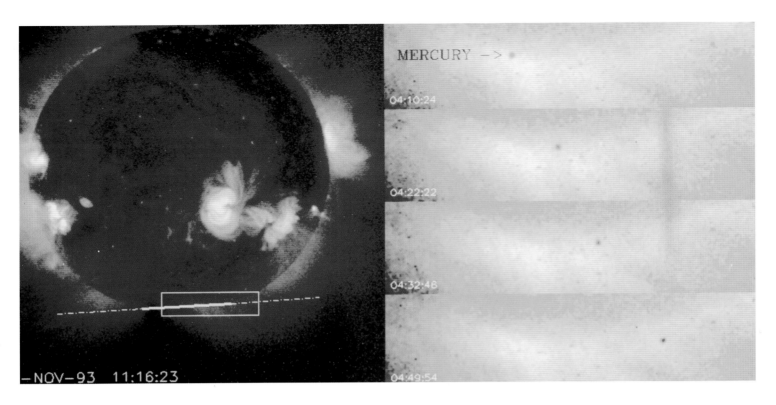

MERCURY —>

04:10:24

04:22:22

04:32:46

-NOV-93 11:16:23

04:49:54

▲ Caught in the glare
*From Earth, Mercury
appears only for a short
while after sunset or before
sunrise; because of the
Sun's glare observers on
Earth can learn little about
the planet. The X-ray image
here, taken from the
Japanese Yohkoh satellite,
in November 1993, shows
Mercury in front of the solar
corona. This was the first
such observation made,
giving useful information on
the still-mysterious corona.*

In the early 1960s, it became known that Mercury was "noisy" at infrared wavelengths; this mystery was explained when the first radar images were obtained in 1965 by the world's largest radio telescope dish at Arecibo in Puerto Rico. When the "echo" of a radar signal aimed at Mercury from Arecibo was detected and analysed, a broadening of the radar signal caused by the Doppler effect (see page 32) was found, indicating that the planet was rotating. It emerged that Mercury took 59 Earth days to complete one leisurely rotation; and the mysterious infrared noise was simply the residual heat that had built up across the planet's surface when it was directly under sunlight, which was radiating away from the night-time side of the planet as it slowly turned.

The data from Arecibo also revealed a connection between Mercury's year and its day. The ratio between them is three to two: the planet spins on its axis three times in the time it takes to complete two orbits. It appears that the immense gravity of the Sun has "locked" Mercury's spin in this way, although why this should be so again remains elusive.

Mariner meets Mercury

To astronomers it was obvious that very little detail of the surface of the world caught permanently in the glare of the Sun would ever be made out from the Earth. It would be necessary to visit Mercury for ourselves. In the late 1960s, JPL planned the first visit to Mercury. As originally outlined, the Mariner 10 mission would have flown past Mercury

◄ Radar eye *In the
1960s, the 300-metre
radio telescope at Arecibo
in Puerto Rico was used to
determine Mercury's rate
of rotation. More recently,
it has been used to image
the planet with radar.
Allowing astronomers to
deduce that, despite the
searingly high temperatures
on the surface of Mercury
that permanently faces the
Sun, water ice may exist in
permanently-dark craters
in the polar regions.*

▼ Plucky Mariner 10
*Because it was operating
near the Sun, the Mariner
10 probe needed only two
solar panels, although a
sunshield had to be added
to protect vulnerable
instruments in the main
body. The craft's main
imaging system was a
TV camera.*

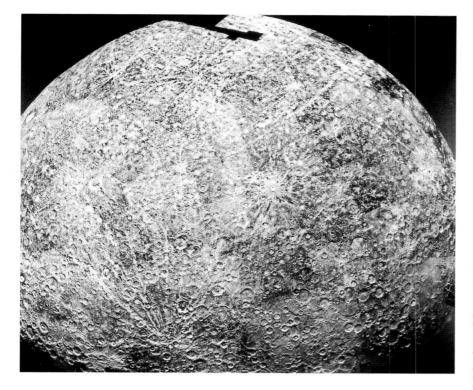

▲ Second pass *Following Mariner 10's second fly-by of the planet in September 1974, this mosaic was assembled from images returned. The pictures revealed the heavily cratered surface of Mercury, which looks superficially like the that of the Moon.*

Because Mariner 10 would be voyaging close to the Sun, the spacecraft's delicate systems had to be shielded by a gold-plated sunshade, and its solar panels had to be tilted to avoid excessive pressure from the solar wind, which would be considerable. The geometry of the fly-past was a compromise, balancing the often-conflicting needs of the various teams of astronomers. In order to map the surface of Mercury, geologists obviously wanted Mariner 10 to pass over the sunlit hemisphere; but to determine whether Mercury had a magnetic field, physicists needed the craft to be sheltered from the Sun, for which it would have to fly past the dark side of the planet. These contradictory needs were resolved by the possibility of making multiple fly-bys and by using a new camera system capable of high-resolution imaging from afar.

Mariner 10 was launched in November 1973, heading first to Venus, which it reached in February 1974, and from which it received a gravity assist towards Mercury. The spacecraft then went on to pass the innermost planet on three occasions: first in March 1974, then in September 1974 and finally in March 1975. Unfortunately, the trajectory of the spacecraft – necessary for mechanical reasons – meant that Mariner's cameras saw the same face of Mercury on each of the three encounters. Shortly after the third fly-by in 1975, contact was lost and Mariner 10 is now destined to orbit the Sun silently for the millennia to come.

just once, after passing Venus en route. However, at a planning meeting for the mission held in 1970, it was pointed out that a special flight path could be chosen that would allow Mariner 10 to intersect the orbit of Mercury repeatedly. By careful synchronization, Mariner 10 could be made to fly past Mercury not once, but at least three times, so providing additional opportunities for gathering data. The idea was accepted, and the necessary fine details of the new trajectory were computed. Yet this glib statement belies the immense complexity of the task involved in executing the necessary gravity assists (see page 36) and in obtaining almost faultless navigation within the limits that were placed on the mission by tracking errors and the need to conserve very limited supplies of fuel.

Mercury's unique terrain

The pictures of Mercury in this chapter represent a tiny portion of the visual legacy of the Mariner 10 mission – there were 12,000 images in all. There was little colour variation in the images because ultraviolet light from the Sun, unfiltered by any atmosphere to speak of, effectively bleached out any differences as seen by the spacecraft's television

▼ First pass *This montage of 18 images from Mariner 10's first pass of Mercury* *was obtained in March 1974. The north pole is at the left, and left of centre is the prominent twin crater Brontë/Degas, connected to other craters by a number of bright rays.*

cameras. At first glance, Mercury appears superficially similar to the Moon – it is covered with heavily cratered areas and lava plains. This suggests that, like the Moon, its surface was heavily bombarded in its earliest epochs, followed by a phase of volcanic activity. On each of its three passes of Mercury, Mariner 10 observed a huge structure, later christened the Caloris Basin, a bullseye impact crater some 1,300 kilometres in diameter, similar in nature to the Mare Orientale on the Moon. The structure has since been filled in by erosion and debris from meteoritic impacts, but the vastness of the initial impact is still readily apparent. So large was the body that excavated Caloris that shock waves were focused on the opposite side of the planet, where they were responsible for forming a strange region of hills and valleys, described as "chaotic" terrain by geologists.

The surface of Mercury was originally molten, and it set – if not in stone, then certainly in basaltic rock – about 1,000 million years after the Solar System's birth. The amount of cratering seen on Mercury roughly matches that on the Moon and Mars, suggesting that there has been little activity since the heaviest period of cratering ceased around 3,000 million years ago. The plains of Mercury are probably the result of volcanic activity, which ended as the planet's internal heat source faded. As the planet's core cooled and shrank, in rather the same way as a piece of old fruit that dries and wrinkles, the surface evolved large cliffs, some more than four kilometres high and hundreds of kilometres in length. Geologists call these features "lobate scarps", and they are unique to Mercury. They are found all across the planet's surface, passing through both cratered and volcanic terrain.

A metallic mystery

After the Earth itself, Mercury is the densest of the Earthlike planets in the Solar System. By measuring the gravitational pull of Mercury on Mariner 10, it was possible for astronomers to determine that while the planet has only 5 per cent of the Earth's mass, it is over five times more dense than water. To explain why a modest body should be so dense, it has been suggested that Mercury must have a large iron core, perhaps accounting for as much as 70 per cent of the planet's total mass. But why might the planet be so rich in iron?

According to the earliest theories, Mercury was believed to have formed in the hottest portion of the swirling proto-planetary dust cloud, close to the Sun. Temperatures would simply have been too hot for rock to have formed and only the heaviest elements coagulated in this proto-Mercury. There is a feeling today, however, that this is probably too

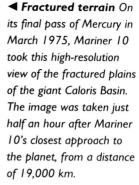

◄ Fractured terrain On its final pass of Mercury in March 1975, Mariner 10 took this high-resolution view of the fractured plains of the giant Caloris Basin. The image was taken just half an hour after Mariner 10's closest approach to the planet, from a distance of 19,000 km.

◄ Lobate scarps Peculiar ridges seem to result from the ancient contraction of Mercury's core and the resultant shrinking of the crust. Craters 30 to 50 km in diameter are visible; some are recent, others were apparently contorted when the ridge was originally formed.

◄ Caloris Basin The largest feature so far seen on Mercury, it is 1,300 km across and was created by a vast impact, the crater of which later filled in.

▼ High-gravity craters Mercury's craters are unlike those on the Moon. Mercury's higher gravity causes ejected material to fall back more quickly.

▼ Oven world This false-colour temperature map of Mercury was made using the Very Large Array (VLA) radio telescope in New Mexico. Red indicates the hottest temperatures; mauve, the coolest. The highest temperatures occur, not surprisingly, on the equator where sunlight is most intense.

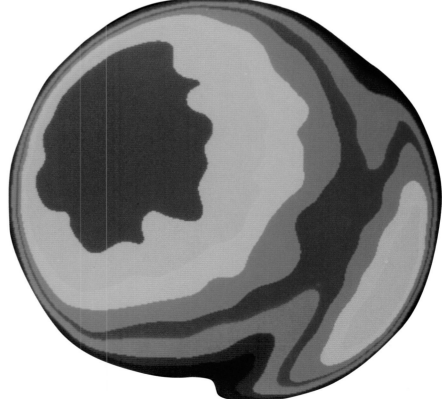

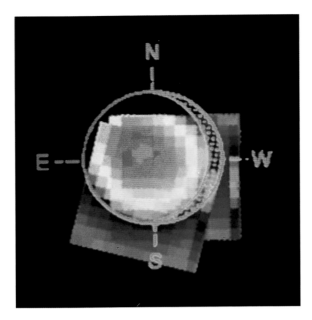

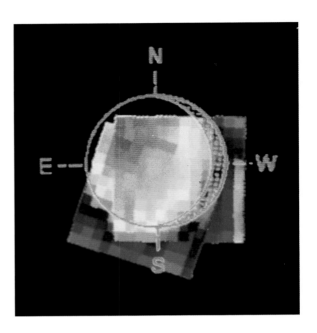

◀▶ **Sodium shift** In the late 1970s, observations from Kitt Peak Observatory in Arizona revealed details of Mercury's magnetic field. False-colour images show Mercury in sunlight (left) and sodium emission (right). The red dotted area is Mercury's night side. The ions of sodium vapour in the atmosphere seem to be shifted towards the north, suggesting a relatively strong magnetic field.

simple a view, and in recent years a rather more complex theory has evolved. This new theory suggests that the original Mercury was twice as big as it is now, and orbiting between the Earth and Mars. The world we see today had its predominantly rocky outer surface layers blasted into space by meteoritic impacts. Either one large body or repeated impacts by smaller bodies were responsible. The Caloris Basin would have resulted from such excavation, which fits neatly with such a scenario, and might also explain how Mercury was displaced into its present highly eccentric orbit.

Given that Mercury is so iron-rich, it is natural for the planet to have a magnetic field. This provides another conundrum to complicate the emerging picture of the innermost planet. In our standard picture of how planetary magnetic fields are generated, a molten core lies within a fairly rapidly rotating body. In the Earth, for example, the daily rotation creates strong swirling motions deep within the molten iron of the Earth's core, and these create the magnetic field, which protects us from solar radiation. Mercury rotates extremely slowly and, based on the geological evidence, seems to have been volcanically dead for many millions of years. Yet paradoxically, Mercury has a comparatively strong magnetic field. One guess is that there may be some impurity in the core that keeps it partially molten at a temperature below the melting point of pure iron.

Hidden icy depths

The most recent news from Mercury has been from the mapping of the hemisphere of Mercury not seen by Mariner 10. (As mentioned earlier, the restrictions placed on Mariner 10 by its flight path meant that the same face of Mercury was seen on each of its three encounters.) The results were achieved by

a development of the radar technique by which the length of Mercury's day was first determined. From August 1991, astronomers at Caltech and JPL beamed radar signals at Mercury from the Deep Space Network antenna at Goldstone in California. The return signal was detected by the Very Large Array in New Mexico, which consists of 27 individual telescope dishes in a Y-shaped arrangement. Radar observations of the previously unseen hemisphere revealed a highly-fractured surface. Features as small as 100 kilometres across could be made out, and at least three large crater basins were seen. Intriguingly, when the huge Caloris Basin came under the radar beam, it was hardly visible. It will be interesting to discover whether the three new crater basins discovered are as physically prominent as their radar signatures would suggest.

One of the more intriguing results of the radar observations concerns Mercury's poles. Mercury's

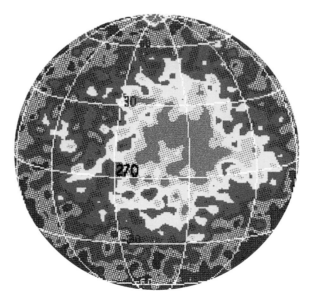

◀ **Icy pole** In August 1991, the VLA was used to detect the echo of a radar pulse beamed to Mercury by a Deep Space Network transmitter in the Mojave Desert. The large red feature at the centre of the image represents highly fractured terrain surrounding an impact basin that was not seen by Mariner 10. The bright reflection at the north pole is believed by some to be water ice at a temperature of about −130°C.

▲ Hot and cold world

An artist's impression of Mercury and the Sun. Because of its closeness to the Sun and its long days – lasting roughly 59 Earth days – Mercury witnesses extremes of temperature from 400°C during the day down to −170°C at night. Some scientists nevertheless believe that there is water ice on the planet in craters at the poles where the Sun never reaches.

orbit is almost perpendicular to its axis of rotation (its axial tilt is just 2 degrees). It is the axial tilt of a planet that causes seasons, first one pole then the other facing more directly towards the Sun as the planet orbits. Thus Mercury has no distinct seasons and the planet's poles do not oblige us by pointing alternately towards the Earth. However, because Mercury's orbit is tilted by 7 degrees to the plane of the Earth's orbit, we are able periodically to get a brief glimpse of the polar regions.

From the reflection of the radar signals and the way they were polarized (the extent to which oscillations in the signal are aligned with one another), astronomers at Cornell University deduced that there may be icy deposits within polar craters on Mercury – material accumulated during cometary impacts in the earliest epochs of the Solar System, just as has been suspected for the Moon (see page 59). The reflection and polarization are characteristic of ice at a temperature of about −130°C. This was somewhat unexpected, because the poles of Mercury are regularly illuminated by the Sun, and such volatile materials should have evaporated long ago. It is possible that, as on the Moon, ice may be hidden from the Sun in deep craters. Clearly, to learn the answer to these mysteries it will once again be necessary to go there for ourselves.

Remarkably, almost everything we know in detail about Mercury was learned by Mariner 10 more than two decades ago. The extreme variation in temperatures on Mercury's surface – a range of nearly 600°C – has more or less ruled out any idea that there will ever be of crewed missions to the inner planet. A return to Mercury was often mooted, but was not deemed scientifically compelling. It is now realized, however, that Mercury holds missing links in our knowledge of the origins of the Solar System. In 1992, NASA's Discovery programme was initiated with the aim of building and launching missions for a total price tag of $150 million; of the ten missions proposed, two concern Mercury.

One is known as the Mercury Polar Fly-by and would specifically study the polar regions and complete the photographic reconnaissance of the planet. It would use an instrument called a neutron spectrometer to detect signs of the hydrogen component of the ice suspected near the planet's poles. Another proposal, termed the Hermes Global Orbiter, would remotely sense the surface, its highly tenuous atmosphere and the all-important magnetosphere. Any future mission will reveal further pieces of the curious puzzle that is Mercury. Only one thing is certain: that the world caught in the glare of the Sun will doubtless surprise us again.

STRANGE SIBLING

Preserved on an ancient fragment of stone from pre-Biblical times is a fitting testimony to the eternal mystique of our neighbour in the Solar System, the planet Venus. Appearing as the Evening Star in the western sky just after the Sun has set, or preceding it before dawn in the east, Venus ingrained itself in the ancient mind for good reason; at times it can be the

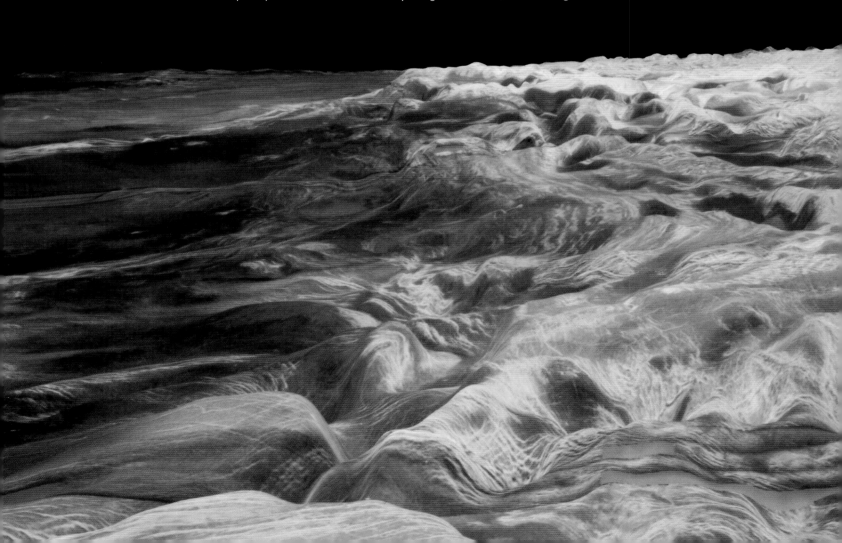

Tortured landscape *The first radar pictures revealed that below the clouds, the surface of Venus was a vision of hell, with twisted contours, complex fault-lines and volcanic fractures. Here a computer-generated view, from radar data collected by NASA's Magellan craft, shows a highland area, Ovda Regio.*

brightest object in the sky, sometimes casting eerie shadows. This mysterious aspect brought Venus to the attention of ancient astronomer-priests. As long ago as 1800 BC, the Babylonians identified the most dazzling of the planets with their goddess of love and war, Ishtar. Part of the text from an ancient fragment of stone reads:

> By causing the heavens to tremble and the Earth to quake,
>
> By the gleam which lightens the sky,
>
> By the blazing fire which rains upon a hostile land,
>
> I am Ishtar.

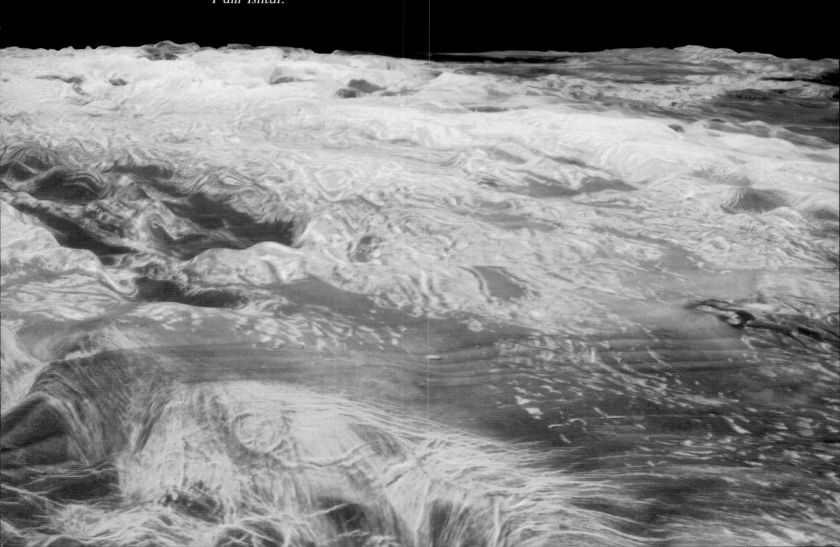

This description of the goddess of love sounds more warlike than one might expect, but it indicates just how much of an impact the planet Venus made. A precedent was set, however, and our planetary neighbour would forever be associated with love. To the Greeks, Ishtar became Aphrodite, the goddess of fertility who had risen from the sea (a notion that inspired Botticelli's painting *The Birth of Venus*). To the Romans, Aphrodite became Venus, originally the goddess of vegetation, but quickly taking on the attributes of the Greek goddess. To the Vikings, Venus became linked with Frigg, the mother goddess and queen of the heavens.

The first astronomical telescopes showed there was more to the dazzling face of Venus: the planet exhibited phases, the waxing and waning of which proved that Venus must orbit within the Earth's orbit, nearer to the Sun. Little else could be discerned about the planet, even as telescopes improved in the eighteenth and nineteenth centuries, and Venus remained a blank inscrutable disk – despite a few fanciful mappings – until very recently. Because the planet was known to be roughly the same size as Earth, astronomers often referred to it as our planetary sister, even though the true nature of the planet and its geography was still entirely mysterious.

When radar technology first became available to astronomers in the 1960s, it uncovered further Venusian mysteries rather than shedding light on existing ones. By measuring the Doppler shift (see page 32) of the surface of Venus, it became clear that our next-door neighbour rotated the "wrong way", and that its day was longer than its year. Why Venus should rotate from east to west (unlike most other planets) and why it takes 243 days to do so (compared with the 225 days it takes to orbit the Sun) remains a mystery still not fully resolved.

The space probe that hit an elephant

The first successful spacecraft to explore another planet was Mariner 2, which flew past Venus in December 1962. There were moments when the probe came close to disaster, but it somehow survived. The early days of the Space Age were plagued by problems with launch vehicles. Mariner 1 had been lost when the navigation system of its Atlas Agena booster failed; the failure was later traced to a hyphen, mistakenly typed in a line of computer code instead of a minus sign. During the launch of Mariner 2, one month later in August 1962, the Atlas first stage unexpectedly started to spin once every

second; but then, by a freakish coincidence, the Agena upper stage malfunctioned and began to counter-rotate obligingly, compensating for the earlier fault. So, as much by luck as by design, Mariner 2 made it to Venus that December, where it effectively rewrote our basic understanding of the planet.

In just 42 minutes of observation, Mariner 2 discovered that the tops of the clouds that shroud Venus were very much hotter than had been expected considering Venus's distance from the Sun. More confusing still, the Venusian clouds had the same temperature in full sunlight as they did on the night side of the planet, showing just how much a dense atmosphere can modify temperatures. Mariner 2 also measured a cross-section through the atmosphere to reveal that, at the surface, the pressure was 90 times greater than on Earth and the temperature was hot enough to melt lead.

Mariner 2 set the trend for subsequent exploration of Venus. Even the best-designed spacecraft would tax controllers to the limits of their ability, and the data returned would continue to challenge astronomers' perceived wisdom about the planet. Throughout the 1960s and 1970s, many more probes were sent to Venus, most of them in the phenomenally successful Venera series launched by the Soviet Union. Starting in 1967, Soviet space scientists designed armoured claddings for their probes, in the hope that they would survive long enough while descending in the corrosive atmosphere to

▲ **Earth's twin** Although in diameter Venus is only 650 km smaller than Earth, geologically speaking it has evolved along very different lines. The Earth's geology is dominated by plate tectonics, and heat from the core escapes to the surface mostly at the edges of the plates into which the crust is divided. With Venus, on the other hand, internal heat seems to be vented much more randomly, although the precise mechanism remains a mystery.

▶ **The best way there** Mariner 2, which became the very first successful planetary mission, at Cape Canaveral just hours before it lifted off for Venus on 27 August 1962. Originally, JPL had wanted to use a more powerful Centaur first stage, but development delays would have meant postponing the launch until 1964. In its place came the Atlas, which had originally been developed for the US Air Force as an intercontinental ballistic missile.

◄ **Mariner 2** *The Mariner 2 spacecraft carried seven instruments to study the radiation and magnetic environment around Venus. The spacecraft design was based on the tapered-tower framework of the Ranger series of lunar explorers. It had a hexagonal box at the tower's base for electronics, and a circular high-gain antenna (at left) for communication with Earth.*

transmit some data before the searing heat and intense pressure took their toll. Thanks to the experience gained, each subsequent Venera probe survived, by and large, for just that little bit longer.

Soviet national pride in its Venus missions and the increasingly political overtones of the Space Race eventually led to one of the strangest confrontations in the history of planetary exploration. In October 1967, Venera 4 survived for long enough to reach the surface – or so it was claimed by the official Soviet news agency, TASS. By comparing the atmospheric pressure profiles obtained from Mariner 5 (the second US spacecraft to Venus, which flew past at roughly the same time as the Russian probe) it was clear that Venera 4 had not, in fact, made it to the surface, surviving only until it reached an altitude of 26 kilometres before presumably imploding. To this clear contradiction from the Americans, Soviet planetary astronomers simply replied that their redoubtable probe had landed on a plateau. American astronomers could hardly believe that their Soviet counterparts would seriously make such a claim – a 26-kilometre-high plateau seemed just a little too improbable. At a later international meeting, a disgruntled Soviet astronomer claimed that stranger things had happened in the past, quoting as an example the fact that the first German bomb to

◄ **Target in sight**
In visible light, the disk of Venus appears featureless. But, as this Pioneer Venus Orbiter image shows, details in the upper cloud deck are revealed with an ultraviolet filter. Although the planet rotates once every 243 Earth days (from right to left as shown here), the cloud tops take only four days for one rotation.

land on Leningrad during the Second World War had killed an elephant in the city zoo.

Soviet credibility was soon restored when a Venera probe did actually land on the surface of Venus and produced some spectacular results. In December 1970, Venera 7 returned data for about 23 minutes, and two years later Venera 8 kept transmitting for 50 minutes, thanks to refrigerated cladding. In October 1975, the next Venera landers

► **Beyond the blue**
A false-colour image of Venus obtained by the ultraviolet spectrometer aboard the Pioneer Venus Orbiter in early 1979. The markings are due to variations in the cloud tops and the presence of sulphur dioxide gas in and above the clouds. An orange-yellow colour has been chosen to show areas where strong light absorption by sulphur dioxide is taking place.

returned the first black-and-white photographs of a rocky surface enveloped in a seething, dense atmosphere. Venera 9 seemed to have landed on a steep mountainous slope, while Venera 10 came to rest on a noticeably stonier surface. Lighting levels, Soviet scientists pointed out, were similar to Moscow on a dark and very cloudy winter's day.

By the end of the 1970s, it was clear that beneath her calm exterior, the goddess of love was more like Dante's vision of hell. The clouds that permanently shroud the surface include layers of pure sulphuric acid. But more than anything else, probes to Venus have revealed that its surface has a remarkable variety of strange geological features, some as seemingly improbable as the chances of that first German bomb killing the elephant in the Leningrad city zoo.

More questions than answers

When the Pioneer Venus Orbiter was launched in May 1978, project officials at the Ames Research Center in California estimated that it would operate for just two years. Few expected that it would still be working more than a decade later. It was the first spacecraft to make radar maps of Venus, revealing an extraordinary landscape of mountains, plains and valleys. The resulting map was on a scale similar to a school atlas of Earth. Eleven other instruments on board scrutinized the atmosphere from the vantage point of a polar orbit; they found little appreciable magnetic field, little temperature difference between day and night, and a collar of cool cloud around the north pole. Although the planet itself rotates very slowly, the clouds race around in the deep atmosphere in only four days, a

◀ **Polar collar** *Infrared observations by Pioneer Venus revealed a vortex of cloud above the Venusian north pole (marked here with a white dot) and a cooler collar of cloud surrounding it (shown in blue). The collar is roughly 35°C colder than the rest of the atmosphere and is another feature unique to Venus.*

"super-rotation" that remains one of the continuing mysteries of the planet. Now atmospheric scientists had the data to begin putting together a picture of how the planet's atmosphere circulates.

For geologists, the Pioneer Venus radar maps raised more questions than they answered. Had there ever been water on Venus? Did the planet show signs of plate tectonics, the process responsible for the Earth's active geology? How old was the surface? There was simply too little detail on the maps to answer these questions. Further maps were made by the Venera 15 and 16 orbiters in 1983, but still greater resolution was needed. In response, NASA planned a follow-up mission in the mid-1980s with a more powerful radar, and named it after the first explorer to circumnavigate the Earth. Launched from the space shuttle in May 1989, Magellan entered orbit around Venus in August 1990, by which time it had come to symbolize the perilous state of the

US space programme. Magellan was assembled from parts left over from earlier space missions — its main radio dish, for example, was a spare from the Voyager mission to the outer planets.

But Magellan was not just a re-run of earlier missions. Whereas Voyager's radio antenna had been used to transmit data and pictures, Magellan used it to map the surface with cloud-penetrating radio beams, using a technique known as synthetic aperture radar (SAR, see page 46). This technique allowed the Magellan dish to imitate a much larger receiver with better resolution, making it possible to see features on Venus as small as 120 metres across. Magellan's orbit was chosen so that it could examine virtually the whole of the planet below, making a long looping orbit over the planet's poles. In the 243 days it takes to Venus complete one rotation, Magellan built up a picture by scanning a narrow strip on the surface under the orbiting probe. This

▲ **The Pioneer Venus Orbiter** *Although planned to last for only two years, the Pioneer Venus Orbiter remained working for 14. As well as new remote-sensing instruments, the Orbiter carried the first radar used to map Venus. In this artist's impression, the magnetometer boom stretches out 5 metres from the main body.*

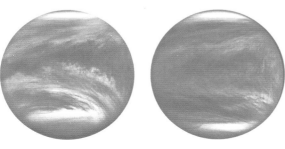

◀ **Y-shaped weather** *An image sequence from 1979 reveals the Y-shaped cloud patterns on Venus. The Venusian atmosphere has only one distinct "cell" from equator to pole, unlike Earth which has three.*

► **First glimpses** *Early radar observations from Pioneer Venus revealed mountainous regions below the clouds. Seen here is a small portion of one of the largest mountains, Maxwell Montes, some 11 km high. Brightness in a radar map does not correspond to brightness as it would appear to our eyes; essentially, darker areas represent rougher terrain, which reflects radar better than smoother land.*

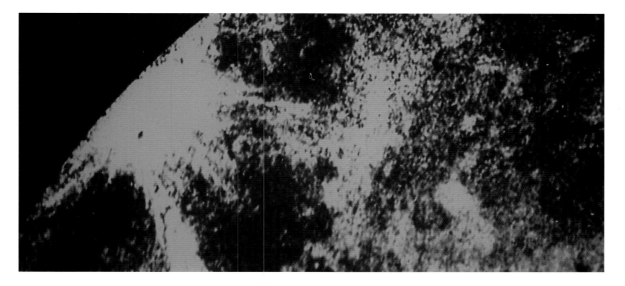

▼ **Drawing a blank**
A colour-coded globe of Venus was generated from Pioneer Venus radar data. Nearly 93 per cent of the surface was mapped, but not the polar regions, which show up here as blank. The highland region coloured green is Aphrodite Terra.

process was likened by one scientist to wrapping a spaghetti noodle around a beach ball 2,000 times.

The Magellan radar-mapping mission began in earnest in September 1990, and over the following two years it recorded 99 per cent of the Venusian surface. The craft completed four distinct mapping cycles: on the first, the radar pointed to the left of the spacecraft; on the second it pointed to the right; the third cycle looked to the left again, but not exactly as before; the fourth was slightly different

again. On each cycle Magellan returned views of the same features from a slightly different perspective, which enabled researchers to produce a remarkable three-dimensional computer model of the surface. This technique of "stereo matching" has meant that subtle geological changes can be discerned. To their surprise, Magellan scientists watched Venus change before their very eyes. Between November 1990 and July 1991, for example, a landslide reduced a 2-kilometre-high cliff to a fan-shaped area of rubble.

► **Mapping complete (nearly)** *A decade after Pioneer Venus, Magellan mapped virtually the entire surface of Venus in far greater detail. A few areas not covered by Magellan have been filled in from* earlier radar mapping missions, including Pioneer and the Venera 15 and 16 probes. This view, from directly over the north pole, reveals a variety of bizarre terrain not seen anywhere before in the Solar System.

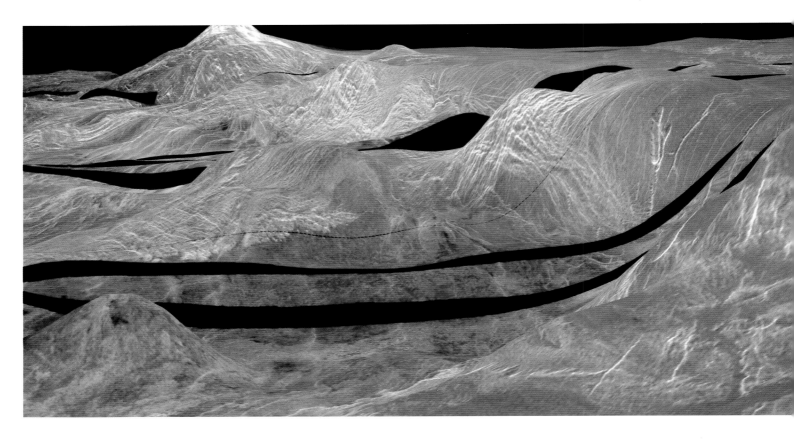

A turbulent past

The highly detailed Magellan maps revealed that the surface of Venus was, in geological terms, remarkably fresh, and that the recent history of Venus was dominated by volcanism. About 85 per cent of Venus is covered in lava flows, including one dried-up lava stream longer than the River Nile. There are volcanoes of all shapes and sizes; the weirdest, which have been christened "arachnoids", are domes a few tens of kilometres across, edged with spider-like ramparts. Even larger are "coronae", volcanic domes that seem to have collapsed – like soufflés – before they solidified. There are also (to continue the culinary metaphor) "pancake" domes, collapsed circular structures, often partly overlapping each other, within which lava has congealed.

There were, however, no tectonic plates. Features seen at the edges of tectonic plates on Earth – scarps, underthrusts and thinning of the crust – have all been seen on Venus, but there seem to be no fault zones or chains of volcanoes resulting from linear movements of the crust. Venus clearly expresses its geological activity in quite a different way to the Earth.

Closer examination of the surface produced a greater, still more baffling puzzle: despite the abundance of volcanic features, Venus seemed, geologically speaking, to be in suspended animation. Magellan's high-resolution images revealed that there are less than a thousand substantial impact craters on Venus (they have been carefully counted and analysed). The craters are spread randomly across the

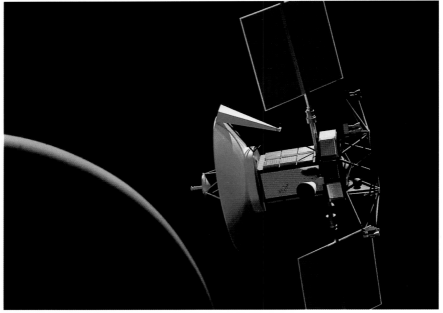

Venusian surface, indicating that all of the underlying terrain soldified at about the same time. Strangely for a planet that seems so volcanically active, the impact craters show little sign of degradation or obliteration resulting from lava flows.

The volcanic surface features on Venus were clearly no more than a few hundred million years old. While the Moon and Mercury preserve a cratering history that goes back almost to the birth of the planets over 4,000 million years ago, Venus's early cratering record has – like the Earth's – been

▲▲ *Strip-by-strip* The *Magellan spacecraft used radar to map the Venusian surface. But gaps were left when data taking had to stop because the dish used to beam radar signals onto Venus doubled as the main communications transmitter. The highland region, Ishtar Terra, is visible here.*

wiped out. It appears that widespread volcanism on Venus ended only 500 million years ago, at which time the planet was almost entirely resurfaced. The craters that are there were created by impacts after the resurfacing phase had taken place. But why did volcanic activity stop so suddenly?

As Venus and the Earth are roughly the same size, they probably have similar inventories of radioactive elements in their cores to provide inter-nal heating. So the two planets ought to generate roughly the same amount of heat internally. On Earth, this heat is lost into space in the form of heat radiation resulting from plate-tectonic movements and volcanic explosions along the plate boundaries. But if Venus has been geologically inactive for 500 million years, how is its internal heat being lost?

Some geologists believe that the global resurfac-ing on Venus may be periodic. The driving force for these episodes would be the gradual build-up of heat beneath the surface crust. When the heat could no longer be contained, a catastrophic event would occur, causing the complete break up of the planet's crust. The whole surface would become a cauldron of molten lava, gradually cooling and solidifying to form an insulating layer, which would again appear inert until the next catastrophic episode occurred.

When the episodic resurfacing theory was first put forward at a conference, many scientists in the audience laughed out loud. According to them, no recurrent geological event can be devastating enough to break up the entire crust of a planet. They claim that craters in the Venusian highlands (poorly seen by Magellan) show signs of erosion,

▶ Pancake domes
Each of these volcanic features is roughly 25 km in diameter, with a distinct pancake-like fractured appearance caused by the collapse and shrinkage of hot lava within them. The structures are located in Alpha Regio in the southern hemisphere of Venus.

▶ Spider volcano *One of many bizarre mountains termed "arachnoids" by geologists was observed by Magellan in March 1992 at Eistla Regio in the southern hemisphere of Venus. Spider-like ramparts surround a slightly concave summit roughly 35 km in diameter. Lava from the central black peak appears to have breached the crater rim at the top of this false-colour computer view.*

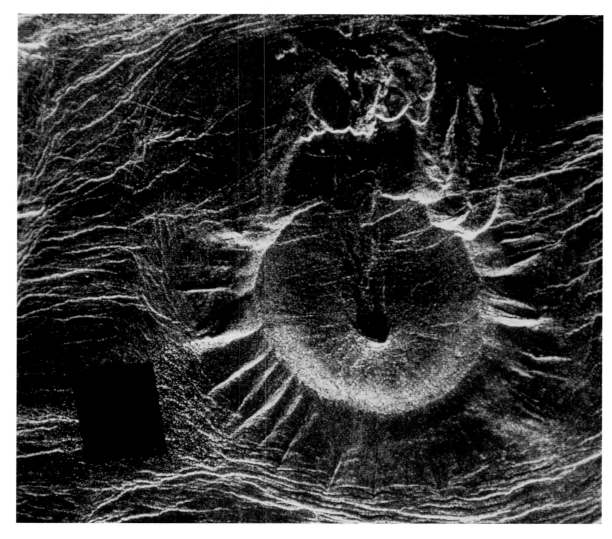

which suggests that they are being resurfaced at a greater pace than the rest of the planet. Some geologists theorize that even higher temperatures in the past kept the surface semi-soft, much like porridge, and as the surface hardened, the imprint of impact craters remained untouched. The landforms seen on Venus today may have been locked in time when the surface became rigid. Regions where the crust is thicker may have stayed warmer for longer and would thus have been active long after the rest of the planet "switched off" geologically.

Other scientists claim that Venus is still volcanically active, but so far any evidence of this has been inconclusive. Earlier probes had detected lightning in the atmosphere and a 15-minute-long thunderclap, which some authorities believe was volcanic in origin. In January 1995 spectroscopic measurements by the Hubble Space Telescope revealed that the atmosphere contained just a tenth of the amount of sulphur dioxide seen in 1978 by the Pioneer Venus Orbiter. Larry Esposito of the University of Colorado believes the sulphur dioxide in the mid 1970s was emitted by a huge volcanic eruption. Since Magellan had observed what could be fresh lava flows around the 9-km-high volcano, Maat Mons, Esposito believes this may have been the source. The sulphur dioxide has gradually been falling out of the atmosphere in the form of sulphuric-acid rain.

An attempt to resolve the issue of volcanism was the last great act of the Magellan mission and

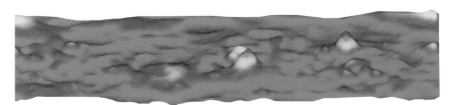

involved the gravity mapping of the Venusian surface. The planet's gravity field should, it was thought, give an indication of how thick the crust is and so how likely it is that episodic resurfacing could take place. The method used to gather data for the map called for some brilliant improvisation.

Any large concentration of material, such as a mountain, alters the strength of the gravitational field above it, and a spacecraft orbiting over it is

▲ Gravity mapping
After its radar mapping, Magellan was used to detect density variations within the Venusian crust. Density changes cause slight changes in the orbital motion of the spacecraft. These three strips show the density variations at the surface (top), within the crust at 15 km depth (middle) and in the mantle at 200 km (bottom). Such observations provide clues to the mysterious evolution of the planet's surface.

◄ Topography and gravity *The gravitational field of Venus (top image), corresponds closely to the visible surface (bottom). The region between Atla Regio (yellow, at left) and Beta Regio (right) has features typical of surface rifting and volcanism. The gravity map suggests that the formation of features was connected with movements of molten material in the mantle.*

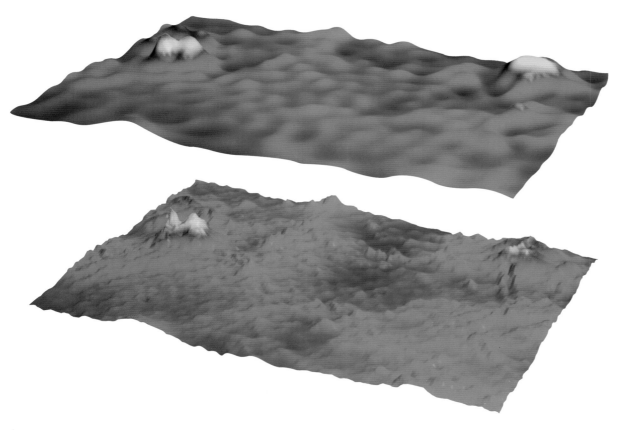

▲ Volcano chain *Three Venusian volcanoes, Ushas Mons, Innini Mons and Hathor Mons, are seen here in a false-colour image of the southern hemisphere.*

The colours are chosen to combine radar "texture" with altimetric height: red and magenta colours indicate the highest areas, and blue the lowest ones.

▼ Surface fractures *Geologists have suggested that the fracture system known as Hyppolyta Linea may result from movements of a tectonic nature.*

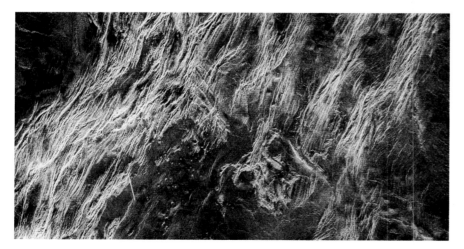

temporarily speeded up by a tiny amount. Similarly, the gravitational pull of any denser area below the crust alters the craft's motion. This speed variation can be measured by the Doppler shift it causes in the spacecraft's radio transmissions. Because the frequency changes are minuscule, it was necessary to bring Magellan in as close as possible to Venus so that the gravity effects would be magnified.

In August 1993, JPL controllers directed Magellan to skim the upper atmosphere of Venus, so that it could be slowed down into a circular orbit closer to the planet. The exercise was hazardous, and many expected it would end with Magellan tumbling out of control into the Venusian atmosphere. But the aerobraking technique worked. In the few months of observation for which funding was available, the so-called "Extended Magellan Mission" returned much useful gravitational data. Scientists are still arguing over the results. Some believe the crust to extend for about a hundred kilometres, which is thin enough to allow for episodic resurfacing. Others maintain it is closer to 300 kilometres thick, which would rule out the possibility.

The new technique of aerobraking is the greatest legacy of Magellan and will, it is hoped, help reduce the cost of future missions to the planets. Instead of using large amounts of fuel to slow themselves into orbit, aerobraking craft will need smaller reserves, although this weight-saving is partly cancelled by the need for protective shields capable of withstanding the fiery heat generated by atmospheric friction.

A pioneer dies

Towards the end of 1992, the Pioneer Venus Orbiter was running on empty and perilously close to the atmosphere of the planet. The craft barely survived by judicious firings of its propellant reserves. However, project scientists at the NASA Ames Research Center managed to learn useful details about the density and structure of the outermost layers of the Venusian atmosphere from the drag taking its toll on the plucky spacecraft. On the afternoon of 9 October 1992, NASA Ames announced that they had lost contact with the probe.

Two years later, it was Magellan's turn. With no further funding for gravity mapping, JPL controllers decided to send the craft in a controlled entry into the Venusian atmosphere. The probe was manoeuvred so that its solar panels experienced atmospheric drag, and the force its thrusters had to exert to prevent the whole craft from spinning was measured. As the craft travelled some 170 kilometres above the surface, it was possible to make detailed observations of the density of the upper atmosphere. This so-called "Windmill Experiment" was the last act of the Magellan mission.

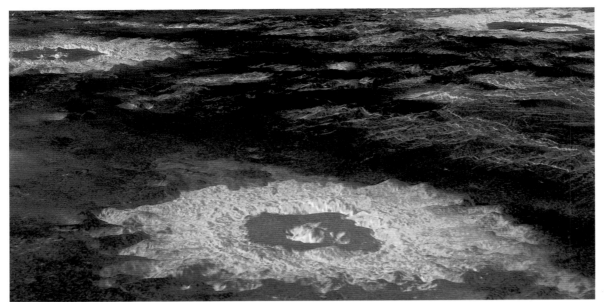

◄ **Craters on Venus**
Despite the young age of
the surface, many impact
craters exist, scattered over
the entire planet. Seen
here are three craters in
the southern hemisphere
observed by Magellan in
September 1991.

▼ **Heat of the clouds**
The yellow clouds in this
1990 Galileo image are 10
to 16 km below the cloud
tops, at temperatures
around 200°C. The blue
and green clouds are higher
up and about 75°C.

A Paradise Lost?

The radioactive dating of isotopes in the Venusian
atmosphere by earlier probes had hinted at the presence of vast reserves of water in the planet's past;
and before Magellan, geologists had hoped to find
ancient sea beds. The relatively young age of the
surface has dashed these hopes. Only a small
amount of water is found in the Venusian atmosphere today. Most of it has long since been broken
up into its constituent hydrogen and oxygen by
ultraviolet light from the Sun (a process known as
photodissociation), allowing the very light hydrogen
to drift off into space. But it is likely that when
they first formed, Venus and the Earth had about
the same inventory of water. Why, then, did the
two planets evolve along such different lines?

The answer may be linked to the abundance of
carbon dioxide in the Venusian atmosphere today.
Most of the Earth's carbon dioxide is locked in car-
bonate rocks and dissolved in the oceans. What happened to cause Venus to release it's store?

In the early Solar System, the Sun was probably
cooler and dimmer than it is now, and Venus was
similar to the Earth, with a reasonably clement climate and with oceans covering much of its surface.
But as the Sun became hotter, Venus's water began
to evaporate into its atmosphere. Water vapour can
be even more effective than carbon dioxide as a
greenhouse gas (see page 43). As temperatures rose
because of greenhouse-type warming, the carbon
dioxide in the oceans was released and entered the
atmosphere to accelerate global warming still further. Then, in a runaway greenhouse effect, the
oceans evaporated entirely. The atmosphere became
the dense, nightmarish place it is today – 90 times
denser than our own, with a temperature at the
Venusian surface exceeding 450°C.

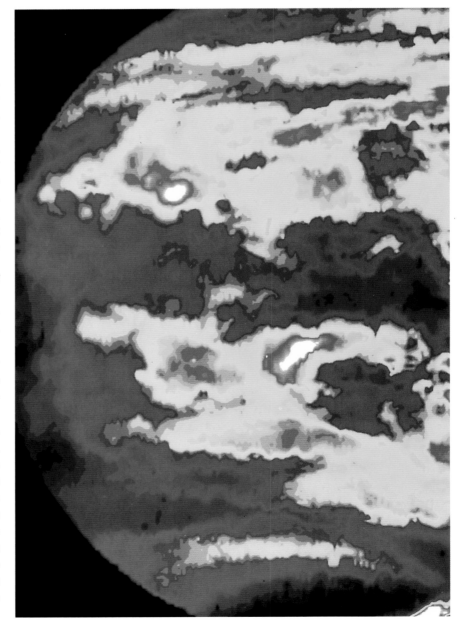

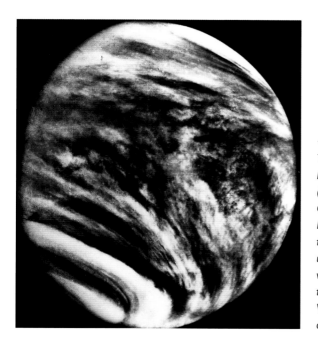

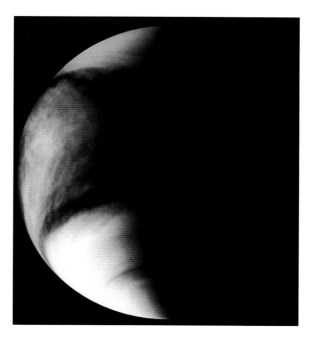

◄► Weather changes
The clouds of Venus as seen by Mariner 10 in 1974 (left) and the Pioneer Venus Orbiter in 1979 (right). Both images were obtained through ultraviolet filters. By using polarimetry, scientists were able to determine that the upper cloud layers of Venus contained droplets of sulphuric acid.

It may well be that Venus truly is a paradise lost, and that only our greater distance from the Sun has saved us from a similar fate. This question is of more than just academic interest, for only time will tell if the rise of carbon dioxide due to human influence will take a similar toll on the Earth itself.

A topography of famous women

The immediate future for the exploration of Venus is NASA's cost-conscious Discovery programme. Of the dozen missions now under consideration, two concern Venus. The first, the Venus Multiprobe Mission, is planned to deposit 14 small probes on Venus to measure atmospheric conditions (an extension of balloon experiments attempted by the Soviet Vega probes in 1985). The second, the Venus Composition Probe, is designed to descend by parachute and resolve the question of how much water was present in the ancient past. Each project is undergoing a three-year feasibility study. These missions will not be launched until well into the next century, but the scientific lure of the brightest planet in the sky is as strong as ever.

Although important questions about Venus remain, the amount of data returned by Magellan exceeds that from all previous planetary missions combined and has rewritten the textbooks on our neighbour in space. In all, around 4,000 features have been added to maps and, as befits the planet of a female deity, they have been named after prominent women. Examples include (Gertrude) Stein, (Simone) de Beauvoir, (Maria) Callas, (Karen) Blixen, (Pearl S.) Buck, and (Margaret) Mead, who all would doubtless have been delighted, as are the scientists who are continuing to examine the geological wonderland unveiled by Magellan.

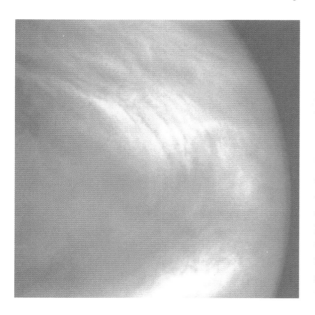

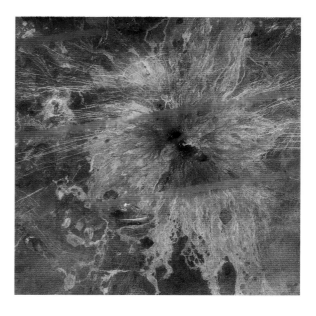

◄ Detail in the clouds
On 12 February 1990, Galileo took this high-resolution image as it sped past Venus. Wave-like structures are visible in the familiar Y-shaped clouds.

► Detail on the surface
Magellan mapped Ushas Mons, a 2-km-high volcano, late in 1992. Bright lava flows and surface fractures are often associated in this type of Venusian volcano, suggesting they formed over "hot spots" in the mantle.

GOD OF WAR

For astronomers, Flagstaff in Arizona is inextricably linked with perhaps the most puzzling and, paradoxically, most familiar planet in the Solar System, Mars. One of the most colourful characters in the study of Mars established an observatory at Flagstaff to indulge his abiding fascination with the Red Planet. Thanks to Percival Lowell, member of a prominent Bostonian family, astronomers were led on one of the greatest of wild-goose chases: the search for canals on Mars. Today, thanks to the efforts of a more recent generation of scientists who also live and work in Flagstaff, we know that, in fact, Martian channel-like features do exist.

A waterless world?
Of our planetary neighbours, the Red Planet, Mars has been explored in the greatest detail and is now largely stripped of its earlier myth and mystery. The apparently green areas, which were once thought to be vegetation, are now known to be of geological origin. However, canals — or at least dried up water channels — have indeed been found. But what has happened to all the water?

The US Geological Survey established its Branch of Astro-geologic Studies at Flagstaff in the late 1960s, and pioneered techniques for investigating the place of terrestrial and lunar rocks in the geology of the Solar System. Flagstaff was an ideal setting, not just because of its association with Lowell. In the nearby Arizona desert stands Meteor Crater, one of the few impact craters on the Earth not to have been erased by weathering and geological activity. Here, planetary geologists had the chance to develop the techniques astronauts used on the Moon, and which robot landers were to use on Mars.

▲ **Mars rock** Although Martian rocks in this Viking 2 image are superficially similar to those on Earth (right), their history and evolution is quite different.

▶ **Arizona meteor crater** One of the few remaining impact sites on Earth, which proved a useful training site for missions to the Moon and to Mars.

Unlike the perpetually cloud-enshrouded surface of Venus, Mars sports features which are easily interpreted by geologists. Mars is the most Earth-like of the planets, and if Venus is the Earth's strange twin sister, then Mars is Earth's diminutive cousin. Although today Mars is an arid, dust-swept, freezing world, it has not always been like this, and the current climate and surface composition of Mars may hold important clues to what the future climate of our own planet might be.

Our current view of Mars is a peaceful contrast to its earlier popular reputation. Appearing as a bright, red-coloured star in the night sky, it was associated by our ancient ancestors with blood and war. The first telescopic observations revealed an ochre-coloured surface marked with darker, apparently green regions, which some astronomers mistook for vegetation. By observing how long it took for these darker regions to move around the planet, it became clear that the Martian day was about 40 minutes longer than our own. Mars has an axial tilt of 24 degrees, almost identical to the Earth's, giving the planet seasons that are visible in the waxing and waning of its polar ice caps.

Mars takes 687 Earth days to complete its oval-shaped orbit of the Sun. At its closest, Mars comes within 207 million kilometres of the Sun; at its furthest, it goes out to nearly 250 million. With just over one-tenth the Earth's mass, and a correspondingly smaller gravity, Mars has been able to hold on

Mariner 6

Mars Observer

▲ **Mars globes on view** Imaging of Mars from space has improved considerably,

although sadly the most recent missions have been failures. In July 1993, the

ill-fated Mars Observer returned only this single black-and-white image,

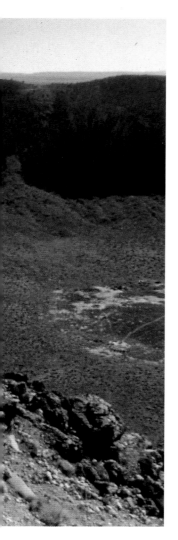

to much less of an atmosphere. Even so, there is enough material in this atmosphere to create violent dust storms. Just after Mars reaches its closest point to the Sun, these storms blow up and often grow to envelop the whole planet in a featureless blanket.

Volcanoes in the haze

After the space age dawned, Mars loomed large in the superpowers' tentative probing of the planets. But whereas the Venera series of probes had made Venus the "Russian planet", fate conspired to make Mars very much "US territory". Good opportunities to launch spacecraft to Mars occur every 25 months, when the relative positions of the Earth and Mars are such that a spacecraft can coast between the two using reasonably limited amounts of fuel.

In November 1964, NASA launched the first successful Mars fly-by mission to return close-up pictures. When Mariner 4 reached Mars in July 1965, its primitive black-and-white cameras returned 11 "usable" frames of the surface as it sped by. Mariner 4 could only transmit data at 8.3 bits per second and, as each frame was made up of 240,000 bits of information, the pictures were tape-recorded on board the craft, and gradually transmitted to Earth over the next ten days, by which time Mariner 4 was long past Mars.

The eleven Mariner 4 images were a revelation. The seventh frame showed something quite unexpected – craters. With the benefit of hindsight after 30 years of planetary exploration, this may not seem at all unusual; but to geologists in 1965, the discovery of Martian craters came as a shock. The craters were reminiscent of the Moon, and to many the Red Planet became the Dead Planet almost overnight. It was a perception confirmed when Mariners 6 and 7 flew past Mars four years later and returned more extensive views of crater-strewn terrain. Both craft established that the Martian atmosphere is composed mostly of carbon dioxide gas, with traces of "inert" gases such as argon. The temperature of the south polar cap suggested that it consists of carbon dioxide ice. Nevertheless, only 20 per cent of the Martian surface was imaged by the three Mariners, and all of that in the southern hemisphere. Further surprises were produced by the first spacecraft to enter orbit around Mars.

When Mariner 9 was launched in late May 1971, planetary scientists were rather more apprehensive than expectant. A few weeks before, its identical twin, Mariner 8, had ended up in the Atlantic Ocean when its Atlas-Centaur launcher had malfunctioned. The burden of mapping the whole Martian surface was to fall to Mariner 9. But as the spacecraft reached Mars that November, Earth-bound astronomers watched the rise of a global dust storm that soon covered the whole of the planet. By the time Mariner 9 entered Martian orbit, astronomers thought there would be nothing to see until the dust cleared. It came as a considerable surprise to see four dark spots – later identified as volcanoes – poking through the haze.

Viking Orbiter

Hubble Space Telescope

from a distance of 5.8 million km. The picture has a similar resolution to the

Mariner 6 image obtained in 1969. The third image was obtained by the Hubble

Space Telescope, before its optical aberration was fixed in December 1993, and the

fourth is a computer-generated view assembled from Viking Orbiter data.

◄ **Orbited and landed**
The last wholly successful US missions to Mars were the two Viking spacecraft launched in 1975. Each Viking consisted of a lander, which descended to the surface, and an orbiter, which examined Mars from space. The Viking 1 orbiter ceased operating on 7 August 1980, the lander on 13 November 1982; the Viking 2 orbiter was switched off on 25 July 1978, and the lander ended transmissions on 12 April 1980.

The discovery of the four large volcanoes was followed in time by a huge valley which, if it was on Earth, would stretch across the entire breadth of the United States. The feature was later named the Valles Marineris, in honour of the Mariner spacecraft. These remarkable features went some way towards reinstating Mars as an exciting place in the popular psyche. The planet revealed by Mariner 9 was a rich, geologically diverse world, shaped by processes apparently similar to those operating on Earth. More interestingly still, the probe discovered features that bore an uncanny similarity to dried-up river beds, hinting that Mars may well have been more Earth-like in the past. In all, Mariner 9 returned more than 7,000 images and gave us our first understanding of Mars as a whole.

The four large Martian volcanoes proved to be of the "shield" type – with broad and gently sloping sides – similar to those found in Hawaii. The largest, suitably christened Olympus Mons after the home of the Greek gods, was 27 kilometres high and almost 700 kilometres across. (For comparison, the largest Hawaiian volcano rises just 8 kilometres from its underwater base.) The three remaining major Martian volcanoes form a closely grouped line across the equator, and rise to nearly the same height as Olympus Mons, but are not as broad. The question everyone asked was why are they so big?

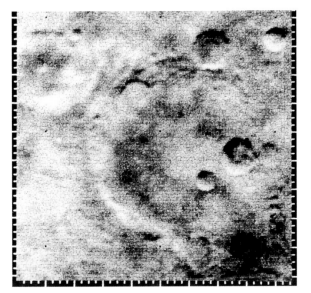

◄ **Frame seven** The seventh frame of the pre-planned sequence of photographic observations from Mariner 4 revealed the unexpected presence of craters in the Atlantis region of Mars. By sheer bad luck, the early Mariners missed the most interesting of the geological features, which are predominantly located in the northern hemisphere of the planet.

The main reason for their huge size seems to be that the crust of Mars has not split into distinct tectonic plates, as the Earth's has done. Volcanoes on Earth, such as the Hawaiian group, were formed by hotspots in the Earth's crust beneath the Pacific plate. They did not reach a great size because the plate gradually drifted across the hotspots, so that instead of a few gigantic volcanoes, there is a chain of smaller extinct ones stretching to Midway Island in the Pacific Ocean. On Mars, the crust is much

▲ Dried-up river bed?
From November 1971 until October 1972, Mariner 9 returned 7,329 images of the Martian surface. The discovery of dried up channels was among the most important revelations of the mission.

thicker and more rigid, so that once a volcano punctured the crust at a single location it just kept on growing. Indeed, the main group of three volcanoes sit on a giant bulge on the Martian equator which itself rises 9 kilometres above the surrounding terrain. Geologists still have not explained how this huge mound – the Tharsis bulge – originated.

At the eastern end of the Tharsis bulge is a network of interconnecting canyons that form the Valles Marineris. The feature consists of many individual valleys, which run parallel to each other before merging into short, broken valleys known as chaotic terrain. The valleys contain areas of landslides and faulting, and are most likely to have resulted from deformations in the planet's crust, perhaps at the time when the Tharsis bulge was uplifted. Out of the eastern end of the Valles Marineris emerge large channels, which were at one time thought to be lava flows; the consensus of opinion now is that these channels were carved by running water. As we shall see later, the atmosphere of Mars today exerts too low a pressure for water to exist on the surface as a liquid, so these channels must be ancient. Many of them run for many hundreds of kilometres, and testify to the fact that Mars must once have been much warmer and wetter.

Viking invasion

In the euphoric aftermath of the startling results of the Mariner 9 mission, it is important to recall that not all Mars probes have been successful. Contact with Mariner 3 was lost soon after its launch in 1964, and Mariner 8 had a watery end; but the Soviets suffered a more constant stream of bad luck. At the time of Mariner 9's arrival in Martian orbit, the Soviet Union attempted to land two probes, Mars 2 and Mars 3, on the surface, but they were soon engulfed by the dust storm. Their landings could not be delayed because they could not be reprogrammed from Earth. Bad systems design, faulty hardware and extreme misfortune continued to plague the Soviet Union's Mars efforts. In 1973–74 four further missions, the Mars 4, 5, 6 and 7 probes, met with varying degrees of failure. Disheartened, the Russians left Mars alone for over a decade, leaving the stage clear for the United States to make the first successful landings on the Red Planet with a pair of Viking spacecraft.

Each of the Viking missions consisted of two parts: a lander and an orbiter. While the lander descended to the surface, the orbiter was to carry out a photographic reconnaissance of Mars from space with greater detail than Mariner 9. After launch in the summer of 1975, the plan was to drop the first of the two identical landers onto the Martian surface on the following Fourth of July, as a perfect climax to America's bicentennial celebrations. The first Viking was to land in a relatively

◄ Atmospheric haze
The Martian atmosphere has less than one per cent of the pressure of Earth's, and is made up mostly of carbon dioxide, with small quantities of inert gases such as argon. One of the largest crater basins, Argyre Planitia, is seen here among the more ancient terrain of the southern hemisphere.

▲▶ Martian Mount Olympus *The largest single volcanic feature in the Solar System is Olympus Mons. The terrain model above gives only a vague idea of its gigantic scale: it spans some 600 km at its base and rises 27 km above the average surface level. The central caldera at the summit is 80 km across, and, as shown in the Viking Orbiter image at right, contains several smaller collapsed craters, probably formed more recently in the history of the planet.*

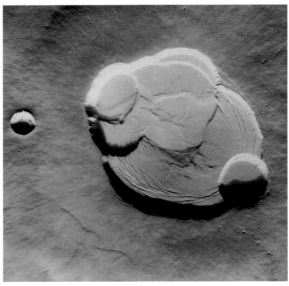

smooth region, where water had probably flowed in the past. Much of the political support behind the mission had been to determine, once and for all, if there were any signs of life on Mars.

When Viking 1 reached Mars in June 1976, its cameras revealed that the planned landing site, which had appeared from Mariner 9 to be a smooth region, was in fact heavily cratered and strewn with rocks. A crash landing on the Fourth of July would have been a dreadful bicentennial present, so an intense effort got under way to find a more suitable landing site. By mid-July bleary-eyed geologists had found somewhere that looked safe, in a region where ancient water flows had apparently fanned out onto the volcanic plains of the northern hemisphere. On 20 July 1976, Viking 1 made a successful landing in this region, known as Chryse Planitia (The Plain of Gold). The craft's cameras revealed the surface to be a rock-strewn desert-scape, in which the sky was brighter than expected and coloured pink by dust particles suspended in the atmosphere. A few weeks later, Viking 2 landed further north in a region called

◄ **Grand Canyon** The Valles Marineris system of canyons runs for 4,500 km with a maximum depth of 7 km. It is seen here in a processed Viking image that simulates the view from 2,500 km above Mars.

► **Cracks at dawn** A section of the widest part of the Valles Marineris. Although Mars is extremely arid, water in the form of vapour or surface frost is often seen, particularly in the early morning.

▼ **Ophir Chasma** A view of the central region of the Valles Marineris canyon system generated from Viking Orbiter observations. The canyons were formed by a combination of faulting and landslides, after which erosion by wind and water took place. The sharp-edged scarp of the canyon wall is roughly 4 km high.

Utopia. It saw a similar landscape, but with rocks pitted with small holes, suggesting they were volcanic in origin as the holes would have been formed by gases escaping as the rocks cooled.

But not as we know it

So was there life on Mars? This question, above all the others addressed by Viking, aroused the greatest interest. Life on Mars has intrigued us for decades. David Bowie sang about it in the early 1970s; in 1938, Orson Welles terrified the United States with his recreation on radio of a Martian invasion from

H.G. Wells's *The War of the Worlds*; and Percival Lowell, more than anyone else, felt certain he had seen direct evidence of it in the 1890s. Today, of course, we know that Lowell's network of canals criss-crossing the Martian surface was illusory; but the discovery by Mariner 9 of dried-up channels with obvious watery origins went some way to suggest that conditions in the Martian past may have been clement enough for life to have evolved.

To see if there were microbes on Mars, each Viking lander was equipped with a fully automated biochemical laboratory. Viking engineers had packed

the apparatus into a volume little bigger than a car battery, and it would be supplied with a soil sample by means of an extendable arm that could scoop surface material from up to 3 metres away. When tested at sites on Earth – including the harsh environment of Antarctica – the laboratory had successfully detected signs of life. The question was, would it do the same on Mars?

The results from the experiment were ambiguous. Many biologists attached to the Viking programme believed that the highly oxidising nature of the surface soil could give spurious results, and furthermore the sampling experiment could only be run once. However, the clincher came from other experiments that analysed the chemical composition of the surface soil; down to one part in a million, these experiments revealed no evidence for the presence of organic material. If there had once been indigenous microbes on Mars, there seemed to be no sign of their biochemical corpses. Some biologists still believe that life may be found at some sites – which they describe as "biological oases"; others think it may even have been driven underground; but the consensus is that neither of these possibilities is likely, and in all probability there is no life of any kind on Mars.

Weather forecasting on Mars

The Viking landers also carried out meteorological observations, acting as automatic weather stations and returning information for many years (Viking 2 operated until 1980, Viking 1 until 1982). At the Martian surface, the atmospheric pressure is less than one-hundredth of that at sea level on Earth, yet this is sufficient for Martian weather systems to have similarities to our own. An obvious missing ingredient is the calming influence of our oceans, but as we shall see, there may once have been extensive seas in the Martian northern hemisphere.

In other ways, though, Martian weather is distinctly peculiar. In the winter hemisphere, for example, the atmospheric pressure drops by 20 per cent as carbon dioxide freezes out onto the pole and the polar cap expands.

Another curious atmospheric phenomenon, the global dust storm, arises in the polar regions. When Mars is at its closest to the Sun, it is also travelling at its fastest. As a result, the southern summer is hotter than the northern summer and the temperature changes more rapidly. At the edge of the southern ice cap, the greater temperature differences during the southern summer often give rise to winds of up to 400 kilometres per hour. These winds generate fierce local dust storms, which circulate throughout the rest of the atmosphere and eventually engulf the planet in a shroud of featureless

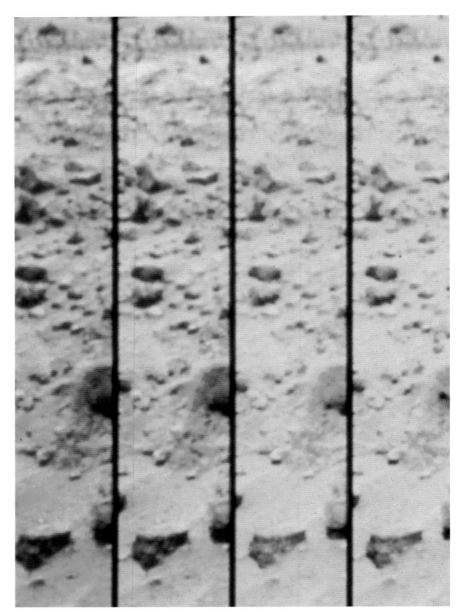

◀ **Peaceful weather**
The Viking landers found that the Martian weather was calmer than expected. High in the atmosphere, winds were known to blow at speeds of up to a few hundred kilometres per hour, but at the Viking lander sites the highest windspeeds recorded were only 20 km per hour. In the thin atmosphere these winds had little force, and few changes were seen in the dust around the rocks on the surface.

▶ **Drifts not dunes**
At the Viking 1 landing site were features similar to sand dunes. In fact, they are not made of sand, but of a much finer dusty material, chemically similar to rust. Martian dust drifts persist for much longer than sand dunes on Earth – despite the occasional Martian wind storm, these drifts changed little during the six years they were observed from Viking 1.

◀ **Swiss cheese on Mars**
Although the two Viking landing sites appear similar, there are subtle differences. This image of Utopia Planitia, where Viking 2 came to rest, shows rocks of volcanic origin. Many of them appear pitted with holes like Swiss cheese (technically "vesiculations"), probably as the result of gases slowly bubbling off as they cooled.

haze for months at a time. The radial wind movements cause the erosion patterns found at the poles to have a distinctly spiral structure.

The peculiar terrain features around the polar regions reveal many clues to past epochs in the Martian climate. Surrounding both poles is "layered terrain", extending for many hundreds of kilometres, in which dust and ice alternate. As we shall see, the amount of dust deposited in these layered terrain regions seems to reveal distinct ice ages in the past, alternating with warmer interludes. Interestingly, another side-effect of the cooler northern summers is that the residual north pole seems to be composed of water ice.

Certainly, water in its liquid form cannot exist on Mars today. Frosts on the surface were often seen by the Viking landers, and fogs and mists were visible by the orbiters usually in depressions like the Valles Marineris. But the temperatures and pres-

sures are such that if you had a pan of water on Mars, it would evaporate explosively. Despite these conditions, the atmosphere often contains water vapour, particularly in the form of clouds over the Martian equator at midday, when solar heating is at its greatest. Clouds have been seen above the Tharsis volcanoes, resulting from the cooling of the Martian air as it is forced upwards around them.

Where did all the water go?

The amount of water vapour present in today's Martian atmosphere is almost negligible. Because Mars formed in a cooler part of the Solar System than the Earth, presumably even more water condensed on it than on the Earth (see pages 23–24). But we can only guess how much water Mars originally possessed, how much has escaped into space because of its relatively weak gravity, and how much remains hidden from view, perhaps bound up

▲ Sunrise in Utopia
In June 1978 the Viking 2 lander recorded this eerie sunrise over Utopia Planitia. From Mars, the Sun is less than half the size as seen from the Earth, though here it appears smaller because of the camera optics.

▶ South pole in winter
Carbon-dioxide ice marks the south polar ice cap, the extent of which changes greatly between summer and winter. The view here is a mosaic assembled from Viking orbiter images obtained in 1980.

▶ Mangala Vallis Close to the equator of Mars lies Mangala Vallis, a region that marks the boundary between the geologically young, volcanic northern plains (shown at bottom here) and the more heavily cratered terrain of the southern hemisphere. Because of its smoothness, and the clear evidence that water at one time flowed in this region, Mangala Vallis is regarded as one of the best future landing sites for any human mission to Mars.

as a deep layer of permafrost below the surface. For water to have flowed across the surface in liquid form at some time in the ancient past, the atmospheric pressure would need to have been just 30 millibars, compared with the average surface pressure today of 7 millibars (the average pressure on Earth at sea level is roughly 1000 millibars).

Can we explain how the Martian atmosphere was warmer and denser in the past? The answer may lie in the water ice deposits and layered terrain at the north pole of Mars. These ancient deposits suggest that major changes have occurred in the orbit of Mars. As we have seen, the Martian orbit is oval-shaped, rather than circular. Careful analysis of the orbit suggest that its eccentricity – a measure of how much it departs from a perfect circle – changes over a period of 100,000 years. When its orbital eccentricity is greater, Mars comes in closer to the Sun and more sunlight reaches the south pole. This, as we have seen, generates fierce storms, which deposit dust on the north polar cap, which in turn creates the peculiar layered terrain.

Similarly, the axial tilt of Mars may have changed by 10 degrees or more over the last hundred thousand to million years (it may have been as high as 46 degrees before the Tharsis bulge was formed in the first 500 million years of the planet's evolution, and "dampened" the axial tilt). When the Martian axis was less tilted, less dust would have been raised, and vice versa. This would have added further to the creation of layered terrain. Finally, the axis of Mars precesses as it orbits the Sun – in the same way that a spinning top wobbles. A complete cycle takes about 175,000 years and the amount of dust deposited will also vary over that period.

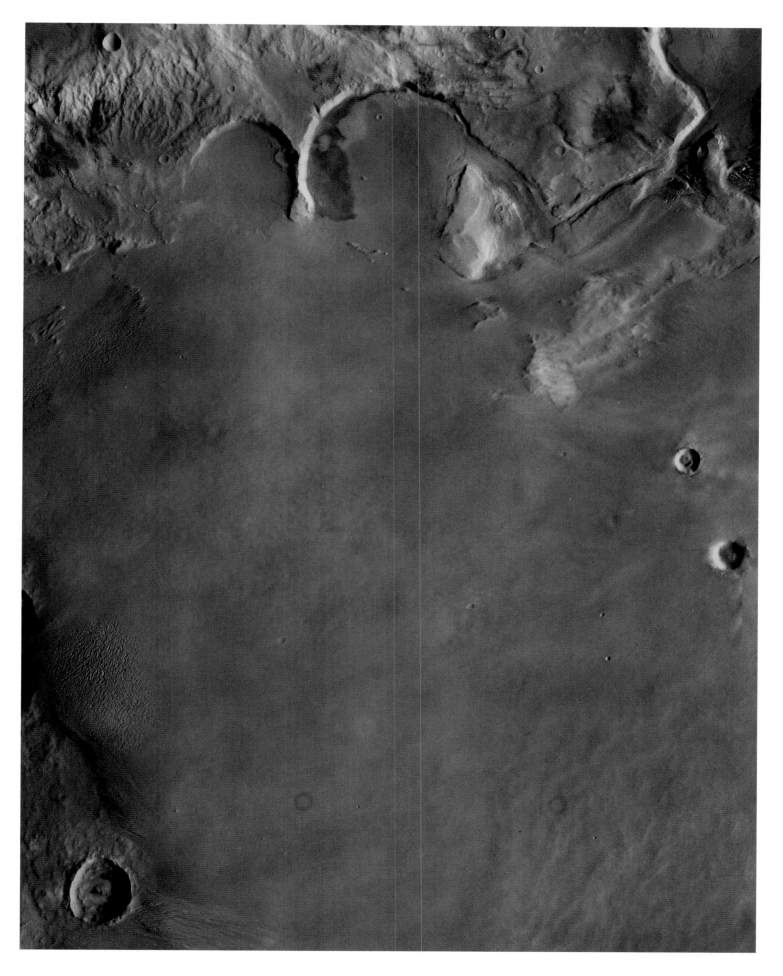

The combined effect of these orbital factors will obviously have influenced the Martian climate. There will have been periods when the planet was warmer and when it was quite possible for water to have existed as a liquid on the surface. What look like sedimentary layers in parts of the Valles Marineris hint that there were once extensive seas and lakes on the Red Planet. To settle this question once and for all, it will be necessary to take core samples from the Martian poles, to determine (as with polar ice cores on Earth) whether the climate was indeed once hotter. If it turns out that these periods coincided with warmer periods on the Earth, it is likely that variations in the output of the Sun played a major role. If that were so, it would help us to understand more about how ice ages are triggered on Earth.

Unanswered questions

In all, the Viking orbiters returned over 52,000 images of Mars and added immeasurably to our knowledge of the Red Planet. And yet fundamental questions about Mars remain. Does it have a magnetic field? What sort of minerals are the different surface features composed of? Why does the southern hemisphere appear so different from the northern and so heavily cratered? These are not merely trivial matters for planetary geologists; they hold important clues to the history of the evolution of Earth. Many of these questions should have been answered by the most recent missions to Mars. Unfortunately, they were not.

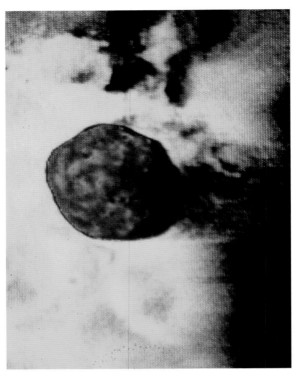

◄ Phobos from Phobos *In February 1989, the Soviet spacecraft Phobos 2 returned this image of the moon after which it was named, against the orange backdrop of the Martian surface below. The craft's camera system returned observations of the moon and planet in the visible and near-infrared regions of the spectrum.*

Sadly, all of the Mars missions since Viking have been heartbreaking failures. In the summer of 1988 the Soviet Union launched two new probes, each to both the Red Planet and its largest moon, Phobos. Named after the moon, the Phobos 1 and Phobos 2 probes should have landed a series of advanced devices on the moon's tiny surface, and returned information about Mars, gathered by a new generation of

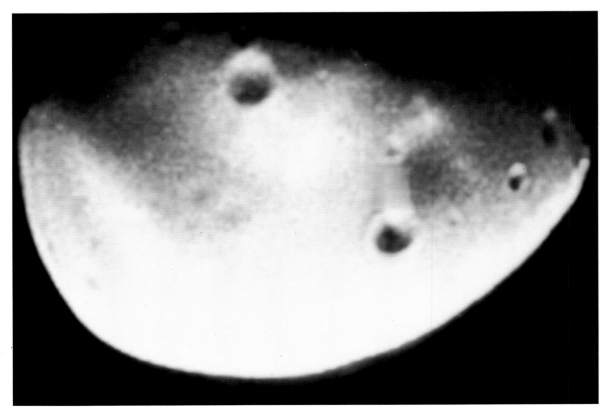

◄ Deimos *The smaller of the two Martian moons, Deimos, is just 15 km across at its widest point. Colour variations in the picture are the result of enhancement by computer; in reality, the surface is a uniform grey-black and as dark as coal. Deimos is about twice as dense as water and, like Phobos, probably consists mainly of a type of rock that makes up certain kinds of meteor. The rock type suggests that it was once an asteroid and was captured by the gravitational influence of Mars. Deimos was discovered – at about the same time as Phobos – by the astronomer Asaph Hall in Washington in 1877.*

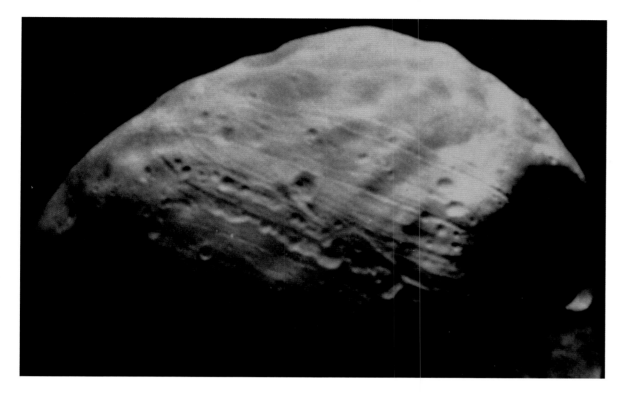

▲ Phobos *The larger of the two moons of Mars, Phobos, measures 28 by 23 by 20 km across. It is dominated by the groove-like "striations" emanating from the 9-km-wide crater Stickney, which itself covers nearly 10 per cent of the moon's surface area. These grooves are either caused by stresses generated by Martian gravity due to the low orbit of Phobos, or are connected with the impact that created the crater.*

remote-sensing instruments. Contact was lost with Phobos 1 just two months after it was launched, whereas Phobos 2 entered orbit and was homing in on its target moon when contact was lost in March 1989. The timing of this last misfortune was especially inopportune as the then newly-elected Council of Deputies was in the process of reviewing Soviet space expenditure, and partly as a result of this failure they reduced it dramatically.

Thankfully, there was at least some scientific return from Phobos 2. By far the most curious result was the discovery that the solar wind may play a far greater role than had been previously believed in stripping away the Martian atmosphere. The solar wind removes an estimated 50,000 tonnes of atmospheric gases each year, a rate sug-

gesting that most of the planet's atmosphere was lost early in the Solar System's history. A related observation concerned the planet's possible magnetic interaction with the Sun. Because Mariner 4 found no sign of a Martian magnetic field, later missions did not carry magnetometers – much to the chagrin of some scientists; the detection of a field, no matter how weak, would reveal much about the planet's interior. The results from Phobos 2, though far from conclusive, suggest that Mars may have a weak magnetic field. The final answer, it was hoped, would be supplied by NASA's first probe to Mars in 17 years, Mars Observer.

In September 1992, engineers at JPL in Pasadena were readying the Mars Observer spacecraft for launch. After Viking, the US planetary exploration

◀ Labyrinth of the Night *In 1989, on the twentieth anniversary of the Apollo 11 moon-landing, President Bush announced that the United States would send astronauts to Mars by the fiftieth anniversary in 2019. This commitment has since been shelved. If astronauts ever do get to the planet, one of the many majestic sights awaiting them is sunrise over the canyons of Noctis Labyrinthus, "Labyrinth of the Night".*

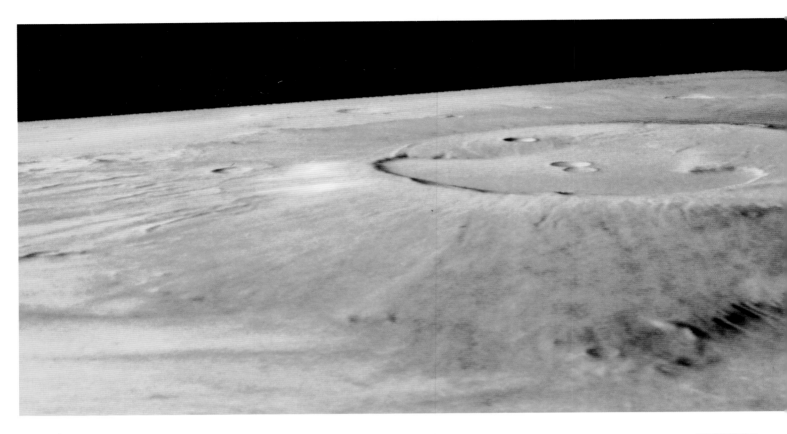

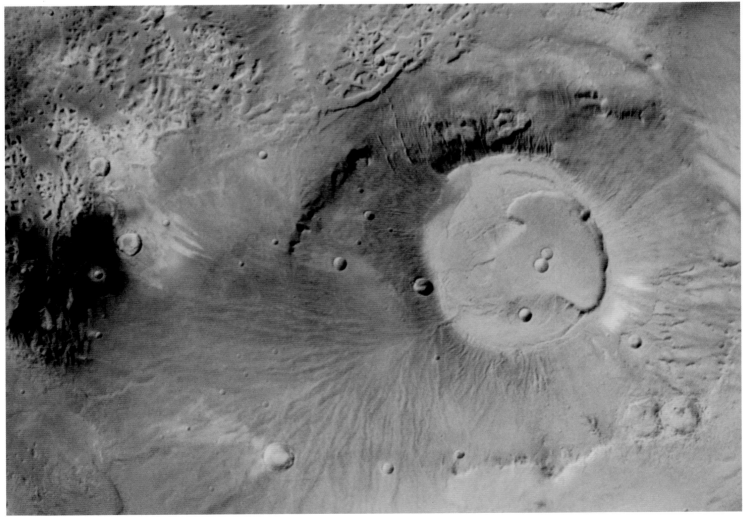

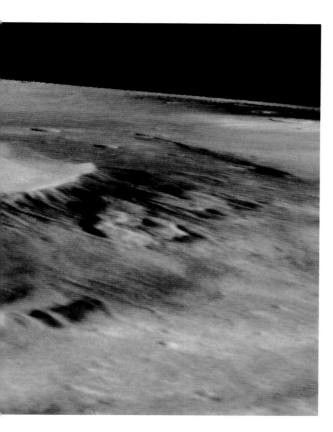

◀ New views from old
The failure of the two Phobos probes and Mars Observer mean there has been little new information from the Red Planet since the early 1980s, when the Viking probes ceased transmitting. But the visual bounty of the Vikings has been transformed in this time with the aid of new computer technology. Mark Robinson, formerly of the University of Hawaii, assembled this startlingly realistic perspective view of Apollinaris Patera from the old Viking measurements and photographs (one of which is shown below).

◀ Apollinaris Patera
As well as the four main "shield" volcanoes on Mars there is another type of volcano called patera. These are ancient, extraordinarily shallow volcanoes, which would not look all that impressive from the surface. Apollinaris Patera is in the southern hemisphere and has a caldera at its summit 100 km wide. The largest volcano of this type is Alba Patera, which stretches some 1,600 km across its base, but is only 6 km high.

programme had been scaled down, so NASA decided to try, as far as was possible, to build a probe to the Red Planet using existing hardware. The mission that became Mars Observer was a modification of an existing design for a weather satellite. The costs of modifying the hardware, however, were probably greater than if they had built it from scratch. The *Challenger* accident (see page 134) had forced an initial postponement of the launch to 1990, and eventually NASA opted for the following launch opportunity in September 1992.

To some little fanfare, Mars Observer lifted off atop a Titan IV booster, the first US planetary mission not to be launched by way of the space shuttle since Pioneer Venus in 1978. Mars Observer was expected to mark the start of a new, confident phase of planetary exploration. But just hours before it was due to enter Mars orbit, on 21 August 1993, all contact was lost. If the spacecraft had merely suffered a computer failure, it would have entered orbit automatically and tried to re-establish contact with Earth (something that the Soviet Phobos craft were not able to do). But no signal was ever found, despite one of the most intense search-and-rescue operations ever mounted by NASA's Deep Space Network. The most likely cause of the craft's failure was that a fuel line burst as its hydrazine tanks were being pressurised in preparation for firing the engines to bring the it into orbit. Mars Observer probably spun out of control with its onboard systems rendered inoperable.

Yet all was not lost. It is standard practice with space missions that, during the manufacture of onboard instruments, flight spares are routinely built at the same time in case there are serious problems before lift-off. Opportunities to launch a mission to Mars occur only every 25 months, and the launch window is only open for a few weeks. It is therefore essential that, if the primary hardware is found to be faulty at the last minute, the flight spare can be substituted rapidly.

After the initial disappointment at the loss of Mars Observer had subsided, it became clear that the flight spares could be used to fit out a complete replacement mission, but this time shared between two separate spacecraft. Duplicates of nearly all the original Mars Observer instruments will be flown, with launches in 1996 and 1998. The first Mars Surveyor craft is to be the larger, and will carry the Mars Observer Camera, which will be able to distinguish features as small as 1.5 metres across on the Martian surface. The second Mars Surveyor craft will carry the remaining instruments.

But by the time the two Mars Surveyors are on their way, Pathfinder, the first American lander mission to Mars for two decades, should already have touched down on the planet. Pathfinder is part of a separate JPL programme that was originally planned as a demonstration for a large and complex programme known as the Mars Environmental Survey (MESUR). Today, in the wake of expensive failures such as Mars Observer (the total cost of which had touched $600 million), "faster, better, cheaper" missions is the new NASA philosophy. Pathfinder is the first in the low-cost Discovery series. With a price tag of only $150 million, the plan is to equip Pathfinder with a "microrover", which will travel over the Martian surface.

At the same time as Pathfinder is launched in November 1996, a Russian Mars orbiter is scheduled to dispatch two surface "penetrators" (with meteorological hardware attached). Originally designated Mars 94, the Russian mission was delayed for financial reasons, as was a follow-on mission. This postponement has actually been of value, because this second Mars probe is now scheduled to carry French-built balloons to be deployed in the Martian atmosphere, as well as a small rover.

For the moment, however, we are left with the legacy of the Viking missions. Using new computerized techniques to generate realistic perspective views of the Martian surface, the original data from Viking has been given a new lease of scientific life. It is these processed views, created by scientists working at the US Geological Survey in Flagstaff, which have created many of the remarkable images accompanying this chapter.

TEMPESTS OF COLOUR

In the early autumn of 1977, the Jet Propulsion Laboratory embarked upon what many scientists regard as the most successful space project ever attempted — the Voyager mission to the outer planets. Up until this point, planetary exploration had revealed vaguely similar landscapes to those of our own world and our Moon, albeit often on much grander scales. But nothing had prepared us for the alien vistas that the two Voyager probes saw in the outer Solar System: the giant gas worlds with their breathtaking weather systems and their vast retinues of moons which were no less diverse and exciting. Voyager 2 would go on to Uranus and Neptune, managing to overcome fearsome technical difficulties. By the time this first survey of the outer Solar System was complete, textbooks had to be rewritten to take into account the wealth of new information.

Turbulent atmosphere
The famous Great Red Spot dominates the rich diversity of features visible in the Jovian atmosphere, which were seen in greatest detail by Voyager 1 in March 1979 and Voyager 2 in July 1979. The vivid colour differences represent clouds at different levels: with the white "zones" higher up than the orange "bands". Jupiter's cloud features intermingle in the most intricate of manners, which are only just beginning to be understood. Some seem to pass through each other without mixing, much like oil and water.

►▼ By name and nature
Pioneers 10 and 11 paved the way to Jupiter and beyond in the early 1970s. Pioneer 10 lifted off on 2 March 1972 from Cape Canaveral atop an Atlas-Centaur launcher. To escape the Sun's influence altogether, Pioneer 10 was accelerated to 52,000 km per hour and crossed the orbit of the Moon 11 hours after launch. The space probe (below) was stabilized by spin, and its simple imaging system built up pictures line by line as the craft rotated. The Pioneers each carried 11 instruments used to measure the magnetic, radiation and dust environment of uncharted space beyond Mars.

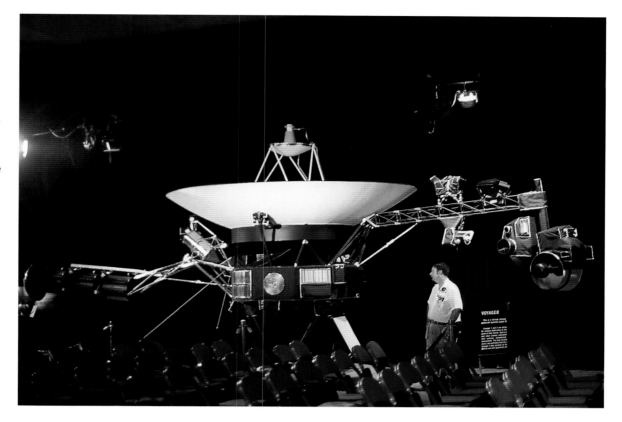

▶ **Voyager** *Far more
sophisticated than Pioneers
10 and 11, the Voyager
spacecraft were stabilised
about three axes, with a
scan platform (at right) for
remote-sensing instruments.
At left is the radioisotope
thermoelectric generator,
the device that provided the
spacecraft with power. The
main radio dish is 366 cm
in diameter.*

The birth of Voyager

The trail to the outer planets was blazed by Pioneers 10 and 11 in the early 1970s. Much simpler than the Voyagers, these probes proved that the asteroid belt posed little danger to spacecraft heading towards Jupiter. This meant that JPL got the go-ahead for a follow-up mission: the two Voyager probes. The Voyagers were to be the best that technology had to offer at the time.

Because in the far reaches of the Solar System, the Sun would be too dim to provide an effective solar-cell power source, the Voyagers were equipped with nuclear power sources. In these the radioactive decay of plutonium was harnessed to provide power. The spacecraft would also have to operate virtually autonomously, for the huge distances involved presented fearsome challenges in telecommunications. Ground receivers had to be upgraded to pick out the faint whisper of spacecraft signals from background radio noise. The Voyagers carried instruments to monitor the solar wind and its interactions with the planets, and a suite of remote-sensing instruments to scrutinize the surfaces of the moons and planets. Last but not least, each Voyager also carried a television camera with both wide- and narrow-angle capabilities.

Both Voyagers were launched in the autumn of 1977. Titan-Centaur boosters, the most powerful rockets then available, were used to achieve the high velocities needed to reach Jupiter. Their trajectories were strictly ordained by celestial mechanics and took advantage of a rare alignment of the outer planets. The technique of gravity assists — where a spacecraft takes gravitational energy from the planet it passes and converts it to additional velocity — would mean the Voyagers would not require impossibly large rocket stages. Both spacecraft suffered problems, as usual, but by far the most serious was the failure, six months into the flight, of Voyager 2's primary radio receiver. Even worse, the back-up system became "tone deaf" and incapable of automatically compensating for the Doppler shift of signals transmitted from Earth. As a result, engineers attached to the Deep Space Network continually

▶ **Interstellar record**
*The two Voyager probes are
now on their way out of the
Solar System. Attached at
the centre of the hexagonal
base of the spacecraft (see
above) each carries the
gold-plated "Interstellar
Record", on which is
recorded sounds and
images of life on Earth.*

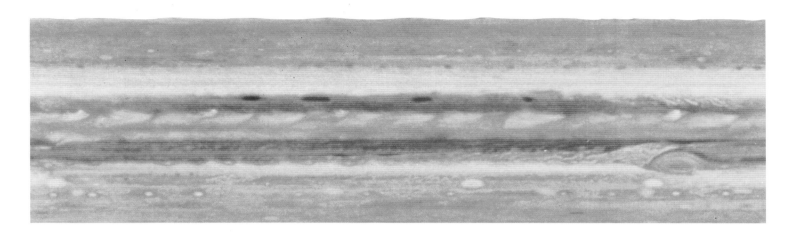

had to calculate the frequency at which Voyager 2 would expect to hear a signal, and then change the transmission frequency accordingly. Initial prognoses of the back-up system's ability to "lock onto" the signal from Earth were not hopeful. It was potentially a major problem an it threatened the longevity of what some already regarded as an ill-starred mission. After such a shaky start, the successes of the Voyagers were all the more astounding. After the first Voyager encounter with Jupiter, one scientist on the imaging team was to remark that the novelty could not have been greater had they encountered another solar system altogether.

Jupiter – a failed star

Jupiter is aptly named after the chief of the gods. Containing about 70 per cent of all the planetary material in the Solar System, Jupiter could hide 1,300 Earth-sized planets inside its gaseous bulk. It is predominantly made up of hydrogen and helium, and theoreticians have calculated that, had it been ten times larger, it would have collapsed to form a star. Indeed, it has been said that Jupiter is best thought of as a failed star, for its core is hotter than the surface of the Sun, and it generates nearly twice as much heat as it receives from the Sun.

Jupiter's interior is thought to be highly exotic. According to the best current picture, a central rock-ice core is surrounded by a vast, deep region of hydrogen which is so dense at its heart, that it conducts electricity as efficiently as a metal does. Jupiter spins extremely rapidly, with a "day" just under ten hours long, resulting in currents in its metallic hydrogen layers that generate a magnetic field some 20,000 times more powerful than the Earth's. This magnetic field gives rise to deadly radiation belts which, if they could be seen with the naked eye from Earth, would appear to be the same size as the full Moon. Above the metallic hydrogen layer is a deep layer of liquid hydrogen, above which the hydrogen thins out to become more like a gas. The atmosphere that we see is the uppermost layer, which is made up of hydrogen gas, with lesser amounts of helium, methane, water, ammonia and hydrogen sulphide. Jupiter's highly colourful appearance comes from complex sulphur and phosphorus compounds that are present in the atmosphere only in minute quantities.

The early images from Pioneers 10 and 11 had hardly prepared scientists for the confusion of colour and incredible wealth of detail in the cloud layers revealed by the Voyagers. So difficult was the

▲▼ *Jupiter unwrapped*
The atmosphere of Jupiter changed visibly between the Voyager 1 encounter in March 1979 (top), and July 1979 (bottom) when Voyager 2 passed the giant planet. The Great Red Spot had drifted westwards, and significant differences were apparent in the cloud flows around it. The belts and zones regularly change in position and brightness, and the atmosphere as a whole seems to go through periods of turbulence and relative quiet. These variations have been studied through ground-based telescopes for more than a century. The recent repair of the Hubble Space Telescope will allow the Jovian weather to be studied systematically without long gaps.

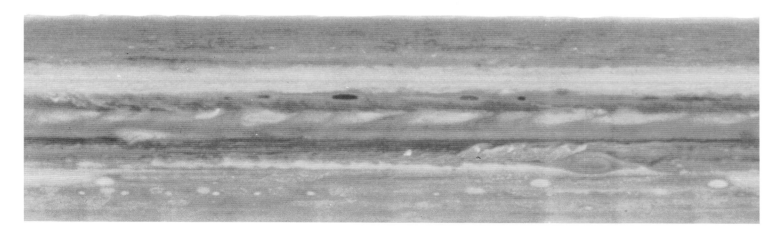

account for all the vagaries of Jovian atmospherics). In the first theory, the temperature difference between the equator and the pole is believed to "drive" the Jovian weather systems, just as on the Earth. Here on Earth, warmer air at the equator rises and heads polewards, and because the Earth is spinning this air flow is spun out into three separate cells, each with its own particular characteristics, such as trade winds. Jupiter also has such cells, but because the planet rotates so rapidly they are pushed out to the top of the atmosphere and produce the alternating belts and zones.

The second, more exotic theory holds that inside Jupiter are vast, concentric cylinders of super-dense, superconducting liquid hydrogen, aligned with the planet's axis of rotation – at right angles to

◄ **Hubble view** On 18 May 1994, the improved second Wide Field and Planetary Camera aboard the Hubble Telescope obtained this image of Jupiter, in more realistic colours than possible from Voyager. The smallest detail is 300 km across – similar to Voyager images from a few weeks before closest encounter. The moon Io casts a sharp shadow.

scientists' task in trying to understand the planet's weather that, at the time of the first Voyager encounter, one scientist remarked that existing theories of Jovian meteorology had been "shot to hell".

Before Voyager, the standard model of Jupiter's atmosphere was that lighter-coloured cloud bands (termed "zones") were the result of rising columns of gas, while darker bands (termed "belts") were caused by descending gas. These, it was believed, were stretched out into alternating, reasonably regular, patterns by the planet's rapid rotation. Observation of cloud motions within zones revealed westerly wind patterns, while the belts were easterlies. At the boundaries between, there seemed to be little evidence for mixing.

The psychedelic maelstrom of colour as seen by the Voyagers called for a dramatic rethink. The boundaries between the belts and zones were not distinct, and large spots (some the size of Earth) swirled within them, sometimes merging, often mingling and then separating. Closer inspection revealed that, although atmospheric motions across the planet were complex, on a smaller scale there seemed to be a degree of order. With all this activity, small wonder that the weather patterns were seen to have changed considerably, not only since the Pioneer fly-bys in the early 1970s, but also between the two Voyager encounters.

Since Voyager, two distinct theories have emerged to explain the nature of the Jovian weather patterns. (They still do not, it must be said,

▼ **Inside Jupiter**
The visible surface is only 1000 km deep. A layer of ammonia crystals (grey-white), are embedded in ammonium sulphide clouds (orange); below this is a layer of water droplets and ice crystal clouds (blue). Internally, Jupiter is dominated by hydrogen.

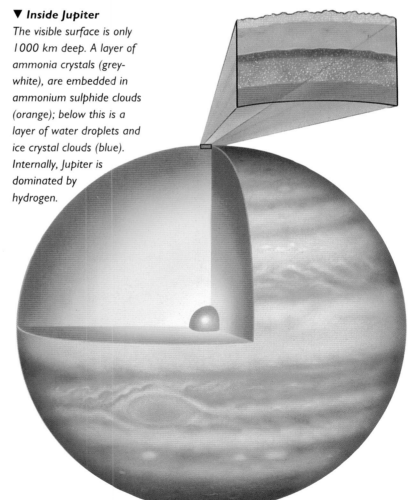

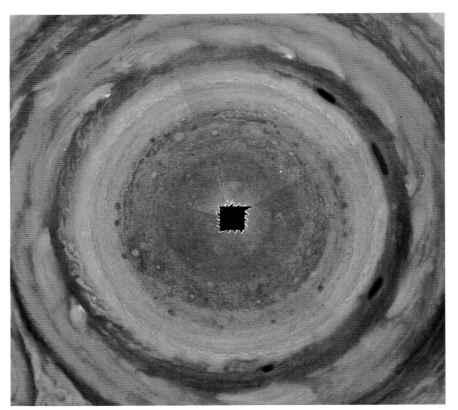

◄ Over the poles
Although neither of the Voyagers flew directly over the north pole of Jupiter, it was possible by computer projection to obtain an image of the polar region. The planet-girdling orange belts and grey-white zones appear to vanish close to the pole itself.

▲ Polar storms *Images from Pioneer 11 offered the first clues to the workings of Jupiter's weather in the polar regions, which are not visible from the Earth. The banded cloud-structure seems to break up closer to the poles, turning into a jumble of hurricane-like convection storms.*

the belts and zones. The belts and zones we see at the surface are actually the tops and bottoms of these "nested" cylinders. This symmetric notion of Jupiter is somewhat spoiled by the presence of the Great Red Spot, a storm system three times the size of Earth, around which wind flows are especially complicated. Clouds outside the spot were observed to be pulled into the spot itself, and then ejected days later. By the time of the second Voyager encounter, a filament of white cloud blocked smaller spots from entering it.

So which theory of Jupiter's interior is the most favoured? Pioneer 11 found that the banding seemed to stop near the poles at 70 degrees latitude, where

the nested cylinders ought to stop. But it was not possible to check this with the higher resolution of the Voyager cameras because the probes made their fly-bys near the Jovian equator. A follow-on mission – then known as the Jupiter Orbiter Probe – would be needed to shed more light on the mystery. Later named the Galileo mission, it was originally intended to reach Jupiter by 1985, although events would conspire to delay its arrival by a decade.

Shock of the new

The naming of a follow-on mission to Jupiter after the great Renaissance astronomer Galileo was indeed appropriate. His observations in 1610, using a simple telescope, revealed four moons orbiting the planet: Io, Europa, Ganymede and Callisto. The discovery of these so called Galilean moons, the first observed in orbit around any planet other than our own, came as a profound shock to the religious world of Galileo's day. For the moons confirmed, beyond all doubt, Copernicus's view that the Earth was not the centre of everything in the Universe.

A similar shock of discovery occurred a few days after the first Voyager encounter when an optical navigation engineer at JPL looked in puzzlement at the screen of her computer terminal. Linda Morabito's job was to ensure that Voyager 1 was on the right flightpath to take it through the Jovian system and onwards to Saturn. Displayed before her was a "raw" black-and-white image of the moon Io, against which a pattern of stars could be seen. She

► Heat map *In the infrared, Jupiter radiates more heat into space than it receives from the Sun. Astronomers used NASA's Infrared Telescope Facility (IRTF) on Mauna Kea to capture this image. Red represents the brightest emission, and the banding corresponds to zones in the upper cloud layers.*

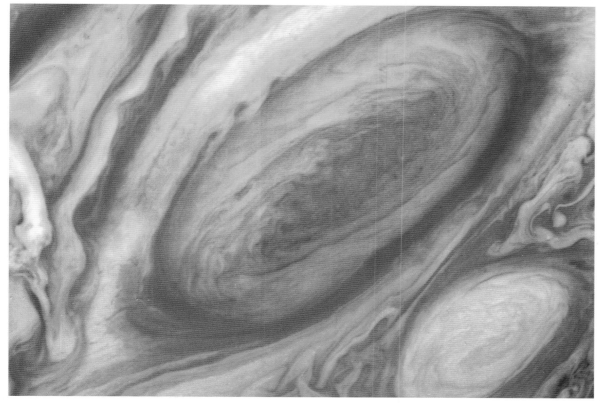

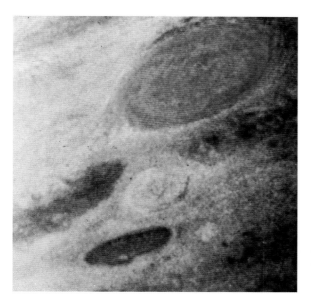

had to compare the actual positions of stars with their computer-predicted positions, and from that determine the spacecraft's location relative to Io. The problem was that there was an unexpected and very bright spot on the moon, and what looked like another moon directly behind it. She asked an imaging team scientist what it might be. It turned out to be an active volcano.

Io was always known to be an unusual world. Orbiting inside Jupiter's lethal radiation belts, it is never going to be a place that humans will be able to visit. Before Voyager, the standard picture of the outer planets' moons was one of inert, icy worlds with crater-pocked surfaces, that had "frozen" in time since the primordial bombardments that accompanied the birth of the Solar System. Io was to be the first of many surprises, for nobody had expected that a world similar in size and density to our own Moon would be the most volcanically active body in the Solar System. The spectacular volcanism explained many earlier riddles: why, for example, a doughnut-shaped "cloud" of atoms and ions (mainly of sodium and sulphur) trailed behind Io in its orbit around Jupiter, and why Io's surface was an unusual orange hue, with a strange infrared spectrum to match.

Io's volcanism is explained by its orbit around Jupiter, for it is caught as a perpetual victim of the gravitational tug-of-war between the giant planet and the next moon out, Europa. Internal stresses, caused by tidal flexing of Io's surface, generate sufficient internal heat to cause the volcanic eruptions seen by the Voyagers. In fact, the eruptions are more like geysers on Earth, except that they spew plumes of sulphurous material hundreds of kilometres above the surface.

In terrestrial geysers, heat from within the crust vaporizes water at the bottom of a column, driving out the water in the form of a hot plume. On Io, the volcanism is powered by liquid sulphur or sulphur dioxide. The surface colouring is a vivid mixture of yellows, reds, oranges and blacks – the colours of the different physical forms (allotropes) of a single chemical element, sulphur. In all, the Voyager craft found nine volcanic vents, although one of these had died down between March 1979 and the second Voyager encounter in July 1979. Together these vents eject an estimated 10,000 million tonnes of sulphurous material per year, depositing a layer roughly a millimetre thick on Io's surface, a rate of resurfacing which, as one Voyager scientist declared, makes Io look only as old as yesterday.

Europa, Ganymede and Callisto

Though Io is by far the most active of the four Galilean moons, the remaining three show evidence of activity that lessens as their orbital distance from Jupiter increases. Jupiter and Io together have a marked effect on Europa, the smallest of the four Galileans. As seen by the Voyagers, Europa bears a passing resemblance to the smooth shell of an egg that has been fractured. In reality, Europa's surface consists of interlocking ice sheets which appear similar to those round Antarctica. Astronomers believe that, under the same tidal forces that keep Io volcanically active, Europa's ice sheets, are continually melting, solidifying and re-melting, giving Europa one of the youngest geological surfaces seen in the Solar System (after Io, of course). This resurfacing means that Europa's surface is almost devoid of craters, and there are no valleys or volcanoes.

▶ **A family of moons**
The four Galilean moons are of startling geological diversity. Here, they are scaled to their correct relative sizes; the moons' diameters are: Io, 3,640 km; Europa, 3,140 km; Ganymede, 5,260 km; and Callisto, 4,800 km.

Our own Moon is slightly larger than Europa, while Ganymede is larger than Mercury. The images of Io and Callisto have been greatly enhanced to bring out details of surface features; in reality, Io is less colourful and Callisto much darker than shown.

Io

Europa

Ganymede

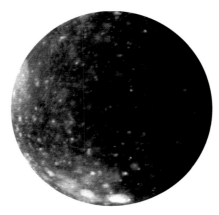

Callisto

Europa's icy crust must be decidedly thin, like the icing on a cake, since the moon has a relatively high overall density. Below the ice there may be a slushy region, kept warm by the tidal interactions with Io and Jupiter. Calculations suggest that if the surface of Europa was molten when it was formed, then there may still be sufficient heat to keep some of the lower surface layers molten. There may even be a subsurface ocean beneath Europa's icy crust.

The outermost Galilean moons, Ganymede and Callisto, are larger than Europa and Io, with lower densities, suggesting that they are almost wholly composed of water ice. Unlike Callisto, Ganymede – the largest moon in the Solar System, and larger even than Mercury – shows signs of internal activity. Both moons are affected very little by tidal forces. Yet Ganymede's surface, with a darker, ancient terrain, is rutted by lighter-coloured regions upon which are icy grooves and wrinkles. It is fair to say there is no agreement on the nature, or origin of these features.

Callisto is the farthest out of the Galilean moons and also the least active. Its outer surface is heavily marked by craters, a record of the heavy bombardment in the early days of the Solar System. Among this ancient terrain fresher, brighter ice is visible within some of the more recent impact craters. The moon has a spectacular bullseye feature, christened Valhalla, which is surrounded by concentric rings. Like the Caloris Basin on Mercury, it is the scar of a particularly devastating impact. Callisto is just what the outer moons of the Solar System were expected to look like before the Voyager encounters.

Galileo – hopes and fears

Spectacular though they may be, the Voyager images are only quick snapshots of Jupiter and its moons – the images are similar in resolution to those of an amateur astronomer taking photographs of our own Moon using a small telescope. Although Voyager discovered a ring around Jupiter, and raised the tally of its moons to 16, very little further information could be obtained about these new objects. Detailed observations will not be long in arriving, however, for the next step in our exploration of Jupiter is finally about to take place.

The Galileo orbiter will arrive at Jupiter in December 1995, and will then release a probe that will descend into the Jovian atmosphere to sample it and examine its structure. The orbiter will then spend nearly two years in a complex orbit, making at least one close pass by each of the Galilean moons. The Voyager images were obtained from many tens of thousands of kilometres away from Jupiter; Galileo will better the resolution of these images by at least one-hundred fold.

That Galileo should be returning information at all is astonishing, for the project has suffered a number of mishaps. The spacecraft had been designed to fly to Jupiter directly, but problems with the space shuttle (culminating in the *Challenger* accident in January 1986) continually delayed its launch. Increased safety fears over the shuttle meant that NASA had to switch the Galileo mission to a conventional launch vehicle. As there was no longer a booster rocket powerful enough to send Galileo on a direct trajectory to Jupiter, the spacecraft's jour-

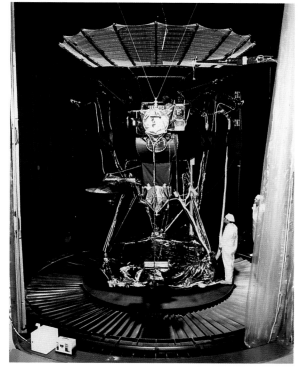

▼ **Galileo** *The most sophisticated planetary spacecraft ever built, is seen here in the "clean room" at the Jet Propulsion Laboratory, undergoing final tests for its long-term reconnaissance of Jupiter and its moons. Galileo was launched in October 1989 by the space shuttle Atlantis, and begins its exploration of Jupiter in December 1995.*

▶ **Dishy problem** *The main dish of the high-gain antenna of the Galileo spacecraft comprises 18 umbrella-like ribs, as seen in this assembly test at JPL. When attempts were made to open the dish in space, at least three of the ribs refused to unfurl. JPL engineers have since been able to work around the problem by condensing Galileo's data signal and by greatly improving the sensitivity of Earth-based receiving stations.*

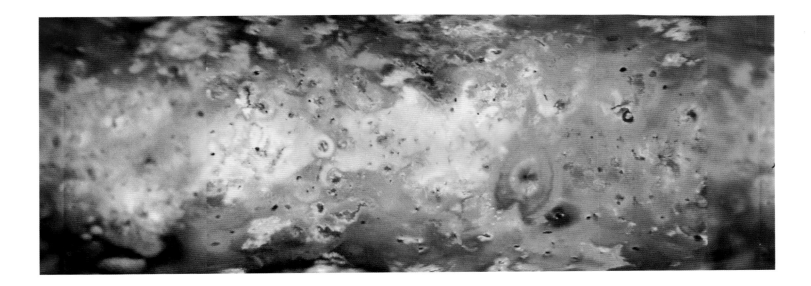

▲▼ The changing face of Io *Differences in the surface of Io between the two Voyager encounters are plain to see in these two Mercator projections. Because Voyager 1 passed closer to the moon, its observations (top) are more detailed than those from Voyager 2 (below). Eruptions from Io's largest volcano, Pele, in the centre right, had altered its "hoofprint" shape. Also, by the time of the second encounter, Pele had ceased erupting. Colours in these images have been enhanced to bring out details.*

ney would now take six years, using gravity assists from Venus and the Earth, instead of about 18 months travelling by the direct route.

After launch by the space shuttle *Atlantis* in October 1989, Galileo began its winding journey to Jupiter by initially heading inwards towards Venus. Protection from the Sun had not been needed for its original, direct route to Jupiter. In the early stages of its new route, Galileo's large radio dish remained furled, protected by a sunshield. The probe communicated with the Deep Space Network by way of a low-gain antenna, a less powerful transmitter that would normally be reserved for routine engineering data. Galileo made a few observations of Venus in passing, and these were stored in the spacecraft's central memory and played back when Galileo came within transmitting range of Earth again. To make room for the data, however, certain stored commands, which engineers thought unlikely ever to be needed, were removed from this memory.

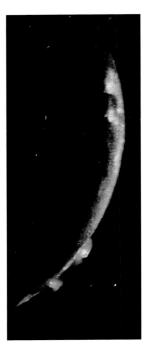

◄ Eruptions in progress *Voyager 2 was specifically programmed to watch for volcanic eruptions on Io after their unexpected discovery four months earlier during the Voyager 1 fly-by. When the crescent form of Io was observed on 9 July 1979, two eruptions were in progress. The predominant blue colour probably results from sunlight scattered by particles of sulphur dioxide.*

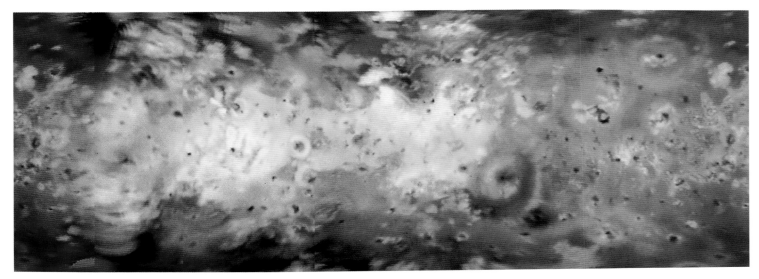

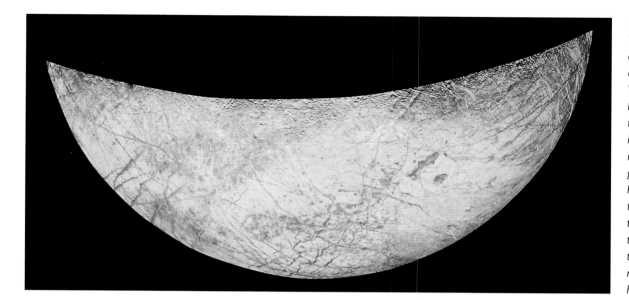

◀ **Iceworld – Europa**
The smooth, icy surface of Europa is seen here in a mosaic assembled from Voyager images. Narrow linear markings criss-cross the moon's icy crust, which is roughly 50 km thick. It is not certain whether these grooved features (which have been compared to car tyre marks on snow) are troughs or ridges, although they differ in height from the surrounding surface by no more than a few hundred metres.

▼ **Ganymede** The large dark area known as Galileo Regio is clearly visible in this Voyager 2 view of the largest moon in the Solar System, Ganymede. Impact craters have exposed the lighter terrain that lay below the dark surface.

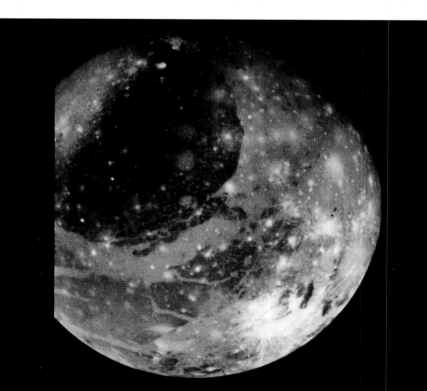

After flying around Venus in February 1990, Galileo passed the Earth in December 1990; this gave the craft the first burst of additional velocity needed to reach Jupiter, and all seemed to be going well. In April 1991, mission controllers at JPL ordered the main radio dish unfurled, as solar overheating was no longer a problem. The dish itself is an umbrella-like structure, nearly 5 metres in diameter, made of a gold-plated mesh. At its base is a small motor which pushes open a series of 18 graphite-epoxy ribs that form the structure of the dish. When fully unfurled, these ribs should slot into position to keep the antenna fully extended. In all, the unfurling sequence should have taken just 3 minutes. But that fateful April morning the motors stalled after just 17 seconds. Despite many later attempts to unfurl the dish, it remained resolutely jammed. JPL engineers believe that three ribs are stuck along the central shaft of the antenna dish. The cause of obstruction is thought to be epoxy resin on one of the ribs, which had either dried unevenly or became dislodged during the long launch delays when the spacecraft had been repeatedly transported between JPL and Cape Canaveral. The software removed earlier to accommodate the Venus data included commands that could have retracted the ribs, enabling the engineers to start all over again.

After Galileo's second pass of the Earth, in December 1992, JPL engineers made their last attempt to open the antenna. Having passed Earth, the probe arced in towards the Sun before heading out again towards Jupiter. The extra heating from the Sun, it was hoped, would expand the central antenna shaft enough to free the ribs. But neither this, nor "hammering" with the drive motor, freed the obstruction, whatever it is, and Galileo's main antenna remains unfurled.

When the antenna problem was first announced to the public, there was gloom in the scientific community, as it was realised that the volume of data transmitted would be cut drastically. The fully deployed antenna had been designed to return data from Galileo's 12 scientific instruments at a rate of 10,000 bits per second. Initial estimates suggested that only one bit per second would be possible, which meant that commands could not be sent out to the spacecraft, nor information returned quickly enough for worthwhile observations to be possible.

But, thanks to developments on the ground, the Galileo mission controllers had a number of aces up their sleeves that go part of the way towards solving the problem. By modifying the way in which Galileo

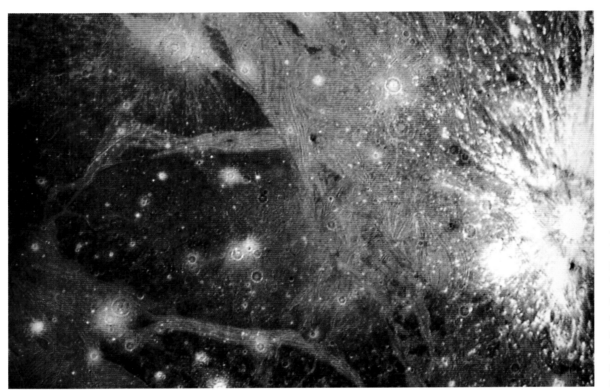

◀ *Groovy Ganymede*
The surface of Ganymede is criss-crossed by lighter "grooved" terrain and darker, older areas that are more heavily cratered. More recent impacts have thrown up fresh, icy material from below the moon's surface.

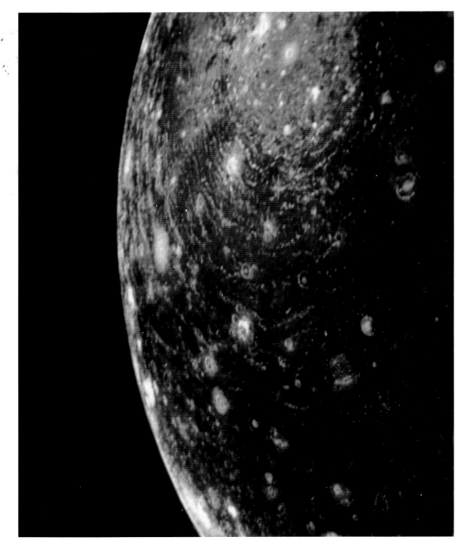

transmits information, the data rate has been increased by compressing it into a simpler form (as we shall see, this is a "trick" that was developed during the later part of the Voyager mission). Improvements to the Deep Space Network also help to "amplify" the rate at which data can be received. More sensitive receivers are available, and groups of radio telescopes on the ground have been electronically arrayed together to pick up the signal more efficiently. In effect, this is like using a larger ear to listen to a fainter whisper.

JPL now believes that Galileo will be able to transmit at the equivalent of 1,000 bits per second from Jupiter. This should allow 70 per cent of the original mission objectives to be met, and more than 50,000 images will be returned. During its two-year sojourn around Jupiter, the spacecraft will make at least one pass, within a few thousand kilometres, of each of the Galilean satellites, as well as making the first examination of most of the smaller moons. Galileo's cameras were tested in 1992 when the craft flew over Antarctica, and fine details of ice flows could be made out. For the Galilean moons, this high-resolution photographic ability, coupled with spectroscopy, should help answer many of the continuing puzzles about their origins.

◀ *Bullseye on Callisto*
On the heavily cratered surface of Callisto lies the vast impact basin known as Valhalla. The central impact area is roughly 600 km across and is surrounded by broken concentric ridges, which extend outwards for some 1,500 km.

▶ **King with a ring** *The two orange lines at the top of the picture were visual evidence of Jupiter's faint ring, as seen by Voyager 2 when it looked back at Jupiter from a distance of 1,450,000 km. The bright bands of colour are the distorted edge of the crescent of Jupiter. The distortion results from the movement of the spacecraft during the long exposures needed to photograph the rings. The rings themselves, seen here from two degrees under the equatorial plane, proved to be much brighter than expected. The left-hand ring image was cut short by Jupiter's shadow.*

LORD OF THE RING WORLDS

The images pouring into the Jet Propulsion Laboratory were nothing short of astonishing. It was November 1980, and Voyager 1's close encounter with Saturn was beginning after more than three years' journey from Earth. For astronomers accustomed to catching fleeting glimpses through telescopes, Saturn was a revelation. The ring system, which had appeared to consist of three distinct bands, was found to be made up of many thousands of rings, and the apparent gaps in the system were found to contain even more rings.

What caused the greatest surprise were phenomena that seemed to defy the laws of physics: dark "spokes" in the rings, and an outermost ring that was perceptibly kinked. Voyager scientists were heard to say that nature did not behave in this way. In time these mysteries would be solved for, thanks to the Voyagers, ring systems were found around all the gas giants and many secrets learned about them. By their brightness and diversity, however, Saturn's rings are by far the most spectacular in the Solar System. More importantly, they give clues about the origin of the planets as a whole and are a benchmark against which our understanding of the other ring systems is now measured.

Majestic rings Saturn's rings are arguably the most splendid sight in the Solar System; they have a width of 270,000 km overall, and are pulled into distinct bands by the gravity of Saturn's moons. The view here was seen by Voyager 2 from 2.1 million km, after it had passed Saturn in August 1981. Such a backlit image of the rings was impossible to obtain except with a space probe, and scientists were all the more pleased to receive it as Voyager 2's scan platform had malfunctioned just days before.

▲▶ Crooked rings?
Two of the more spectacular discoveries made by the Voyagers were the braided appearance of Saturn's outer F Ring (above), and the "spokes" within the central B Ring (right). At the time, these phenomena seemed to defy the laws of nature, although they have since been explained.

The riddle of the rings

That the rings of Saturn proved so puzzling was nothing new. Galileo had been baffled by them when he made his first telescopic observations of the planet in 1610. Saturn was then assumed to be the outermost of the five known planets, wandering slowly through the skies, taking nearly 30 years to orbit the Sun. At an average distance of 1,427 million kilometres from the Sun (twice as far as Jupiter), Saturn marked the edge of the Solar System, until the discovery of Uranus in 1781.

Through the first astronomical telescopes, Saturn looked quite bizarre. There appeared to be two smaller objects either side of it, which caused Galileo no end of anguish – especially when he observed Saturn two years later and they seemed to have disappeared. The primitive nature of his telescope was such that Galileo could not resolve the puzzle of the rings, and it was not until 1655 that Christiaan Huygens saw them clearly and interpreted them for what they were. In the same year, Huygens observed the first of Saturn's nine moons to be discovered before the Space Age – the giant moon Titan, second only to Jupiter's moon Ganymede in size and larger than Mercury.

In time, improved telescopes revealed three distinct "bands" in the ring system, named A, B and C, with the C Ring closest to the planet. There was a prominent gap between the A and B Rings, which was named the Cassini Division, after Giovanni Cassini who had first observed it in 1675. It was

clear that Saturn experienced seasons, for the rings alternately tilted north and south by 27 degrees away from the horizontal. Careful measurement had shown that the rings stretched out to more than

▲ Rings against Saturn
The complex nature of the ring system is evident from this view of their outermost portions seen against the backdrop of Saturn. The strange "O" shapes on the atmosphere are caused by dust on the camera.

◄ Fine structure The narrow F Ring was itself found to contain many discrete ringlets, as this synthesized Voyager 2 photograph reveals.

135,000 kilometres from the centre of Saturn, but could be no more than a few kilometres thick. For when the plane of the rings passes through our line of sight, they vanish — a phenomenon that caused Galileo much consternation. Analysis by spectroscope showed the rings to be bright and icy, although there was no way they could be solid: by the laws of orbital motion, material near the inner edge of the rings had to be travelling around Saturn faster than material farther out, so a solid ring would be torn apart (or, more to the point, would not have formed in the first place). The rings had to be composed of separate chunks of ice.

That, in a nutshell, was more or less the extent of our knowledge of Saturn before the Voyager encounters. There were hints that Saturn would

turn out to be even more perplexing than Jupiter, thanks to an earlier spacecraft fly-by. In September 1979, Pioneer 11 made a close pass of the ringed planet, after drastic re-routing following its encounter with Jupiter nearly five years earlier. Although primitively equipped, Pioneer 11 was able to measure Saturn's extensive magnetic field. At 1,000 times stronger than the Earth's, this magnetic field was extremely puzzling, for its axis almost coincided with Saturn's axis of rotation. (Theorists had expected the field to be asymmetric, the result of uneven currents sloshing deep within Saturn's core.) Another Pioneer 11 discovery was a very thin outermost ring and a new small moon, both detected from the "wake" they made in Saturn's magnetic field. And, living up to its name, Pioneer 11

broke further new ground by showing that a space-craft could survive passing through the plane of Saturn's rings without harm.

Just over a year later, Voyager 1 discovered something even stranger, just six weeks before its close pass of Saturn. Its cameras returned images showing dark, radial streaks in the rings, reminiscent of the spokes of wheels, which kept their shape as the rings rotated. Physicists believed such structures should not exist, as they ought to have been distorted by the rotation. The answer to this riddle would come later, once the full range of Voyager observations were examined together.

The rings are indeed made up of countless millions of icy chunks, most of which were far too small to be resolved by the Voyager cameras. The ring pieces come in a wide range of sizes — from as large as a house, to as small as a snowflake. This icy material is marshalled into thousands of individual ringlets, some oval rather than circular, and others seeming to spiral in towards the planet. Computer

▲ **Technicolour rings**
Saturn's rings are mainly water ice, but small colour variations (greatly enhanced in this Voyager 2 image) may result from subtle differences in composition, heightened by light scattering effects. The smallest features that can be distinguished here are about 20 km across.

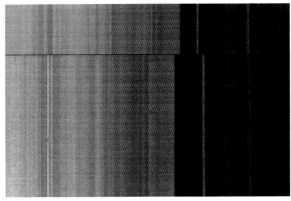

modelling suggests that the spiral form results from periodic gravitational disturbances caused by Saturn's inner satellites, with most ringlets occurring at the crests or troughs of "density waves". The explanation for the dark spokes comes from Saturn's magnetic field, which levitates microscopic ice crystals out of the main ring plane, causing their shadows to fall on the rings below.

▲ **Imperfect circle** If the rings on opposite sides of Saturn are compared, the fine structure does not quite match – the rings are not exactly circular. The pull of Mimas and dynamic waves are the likely causes.

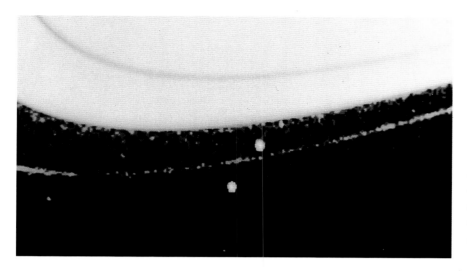

◀ **Prometheus and Pandora** *The puzzling kinked nature of the F Ring is explained by the presence of two small "shepherd" moons, christened Pandora and Prometheus. They are irregular, icy bodies about 100 km across.*

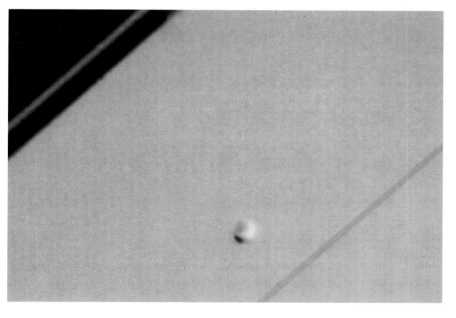

▲ **A shepherd in butterscotch** *On the day of its closest encounter with Saturn in 1981, Voyager 2 returned this remarkable picture of the thin F Ring (bottom right) and a shepherd moon against the butterscotch-coloured cloud tops of the planet below.*

Before Voyager, Saturn's rings were generally regarded as primordial material left over from the planet's formation – pristine ice in deep freeze. This material, it was believed, had been prevented from accreting to form a moon because of Saturn's overwhelming tidal forces. At the distance of the rings, anything object a few hundred kilometres or so across would have been ripped apart by the tremendous gravity of the planet. But as we shall see, thanks to Voyager we now know that all planetary rings in the outer Solar System are – in geological terms – relatively young.

The shepherd's tale

Saturn's rings are much brighter than the ring systems Voyager 2 was to find at Uranus and Neptune. Because the ice making up Saturn's rings is so bright, it cannot have been around for long enough to have darkened to the same extent as the rings of Uranus and Neptune. It is likely that Saturn's rings are only about one hundred million years old.

There is enough material within the rings to make up a small, icy body, a few hundred kilometres across. If such a body ever existed, how it came to be pulled in towards Saturn and ripped apart remains a puzzle that can only be solved by measurements from future space missions.

Perhaps the strangest individual ring was the thin F Ring, discovered by Pioneer 11 in 1979. It lies beyond the outer edge of the main ring system, where it was once believed material could not persist for long, as it was thought this material would slowly diffuse in towards the planet. On close examination, the F Ring in some places appeared kinked and in others it seemed braided. Although to begin with these observations of the F Ring merely served to confuse matters, in the end they were to provide a vital clue in solving how the rings are kept together.

Systematic observations revealed the presence of two tiny moons either side of the F Ring. It turned out that this ring is continuously changing its shape under the gravitational influence of these moons. The one nearer to Saturn travels faster, and tends to pull ring material in towards the planet; the outer moon tends to drag ring material outwards. The gravitational yin and yang from these moons sorts the F Ring particles into different strands according to their size. The extent of the braiding depends on the positions of these so-called "shepherd moons" in their orbits, as the braiding is less marked in material farther away from them.

Theorists were once confident that they understood the workings of planetary rings. Today, more than a decade after Saturn's rings were photographed by Voyager, they are less sure. Before Voyager, the many distinct gaps in the rings had been explained by gravitational "resonances" with the larger inner moons, way outside the ring system. The most obvious gap, the Cassini Division, was thought to result from the influence of the

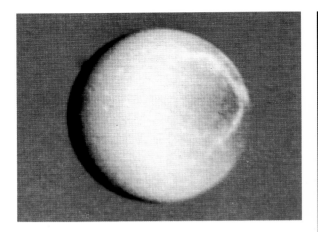

▲ **Iceball** Dione is an icy, bright moon, 1,100 km in diameter, with light streaks across a darker terrain. Saturn forms the backdrop in this Voyager 1 image.

moon Mimas, whose orbital period is almost exactly twice that of material within the Division. Such regular gravitational perturbations would, theorists suggested, eventually sweep the region clear of material. Yet the Voyagers found ringlets within the Division, and it became obvious that there was more than simple resonance at work.

After Voyager 1's discovery of the first shepherd moons, Voyager 2 was reprogrammed to search for more such objects embedded within the main body of the rings. Originally, none were found, but detailed examination of a portion of the outer A Ring showed its edges to be wavy, a tell-tale sign of recent interaction with a shepherd moon. About ten years after the Voyager 1 encounter, further enhancement of the its pictures revealed the moon in question, exactly where calculations predicted it would be. It remains for a follow-on mission to look systematically for evidence of shepherd moons in the other ringlets. Further clues will be sought by the Cassini/Huygens spacecraft, due to arrive in orbit around Saturn in June 2004.

Ice worlds

The Voyager encounters took the total number of Saturnian moons to 18, most of which are small irregularly shaped bodies. The seven sizeable moons have densities one to two times that of water ice, indicating that they must contain some rock. One of the more intriguing of the moons is Mimas, which bears a giant impact crater one-third of the moon's diameter. This, and the smaller craters, are a clear sign of the intense cratering period in the early years of the Solar System. How Mimas survived such a catastrophic impact is not known for certain, but it may have to do with the peculiar way that ice behaves in the freezing cold of the outer Solar System. Saturn's other sizeable moons also have their share of craters. Some additionally have brightly streaked surfaces, caused perhaps by ice melting as a result of cratering impacts, or possibly by some form of geological activity.

The most striking evidence for geological activity is seen on Enceladus, oddly enough one of the smaller of the seven larger moons. Enceladus was

▲ **Mimas** The pockmarked face of Mimas is typical of the larger moons of Saturn. The heavy cratering of the surface suggests that little geological activity has occurred in the aeons since the end of primordial bombardment when the Solar System was young. The brown coloration is consistent with the dirty ice composition suggested by density measurements. Features as small as 2 km can be discerned in this Voyager 1 image.

▶▶ **Enceladus** An unexpected highlight of the Voyager 2 encounter with Saturn was provided by Enceladus, which had been poorly seen from Voyager 1. On the morning of closest approach, 25 August 1981, the new high resolution images caused considerable excitement in the JPL Press Room. Enceladus has an exceptionally bright surface dominated by water ice, at a temperature of –200°C.

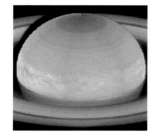 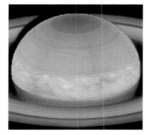 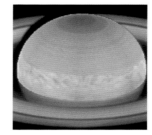 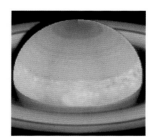

► Storms on Saturn

Storms in the atmosphere of Saturn are relatively rare. A spate of activity was observed by the Hubble Space Telescope in November 1990. This sequence of enhanced images shows the rapid evolution of the storm over just a few hours.

► Bumblebee shots

In the nine months between the two Voyager encounters, Saturn's belt structure changed considerably, as can be seen in these greatly enhanced images. At the time of the second encounter (left), the rings were brighter because they were tilted more directly towards the Sun.

photographed only poorly by Voyager 1, but seen in greater detail by Voyager 2. Groove-like terrain cuts across a more heavily cratered, ancient surface, vaguely reminiscent of Jupiter's largest moon, Ganymede. But with an average diameter of only 500 km, Enceladus is much smaller, and some theorists have suggested that tidal action may be the cause of its internal heating. However, such heating ought to be stronger on Mimas, which is closer to Saturn. Yet Mimas shows few signs of geological activity, and so the problem remains unsolved.

Perhaps the most peculiar of Saturn's moons is icy Iapetus. Ground-based observations had shown that one hemisphere is considerably brighter than the other and that, like the Earth's Moon, the spin of Iapetus is synchronized with its orbit around Saturn. Voyager images confirmed that one hemisphere is indeed very bright, predominantly covered with water ice, while the other is almost black. The 1986 fly-bys of Halley's Comet have shown that cometary nuclei are very dark, so dark surfaces in themselves are not unusual. But quite why there should be such a contrast between the two hemispheres of Iapetus remains a puzzle.

Butterscotch planet

In all the excitement over the rings and moons, the surface of Saturn itself was largely overlooked. After the psychedelic hues and swirling cloud patterns of Jupiter, it is fair to say that Saturn paled in comparison. Seen from Earth, Saturn's atmosphere showed evidence of banding but, as Voyager approached, Saturn remained an almost monotonous butterscotch colour. A few cloud patterns were seen, but Voyager scientists were really hoping for a storm to develop, of the kind that had occasionally been observed from Earth.

Saturn is, in many ways, a scaled-down version of Jupiter, so scientists did expect that its atmosphere would not be quite as dynamic. Yet there are a number of startling differences. Saturn's density is so low (almost half that of Jupiter), that it has been said that if it were possible to find a large enough expanse of water, Saturn would float. The planet is also distinctly oval in shape, bulging perceptibly at the equator, because of its extremely rapid rotation. Saturn has the second-shortest "day" of all the planets, at only 10 hours 14 minutes at the equator, and 10 hours 40 minutes deeper down in the planet.

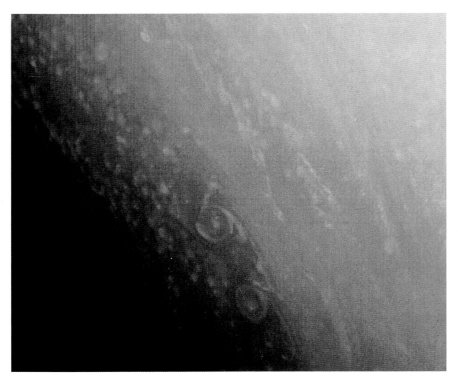

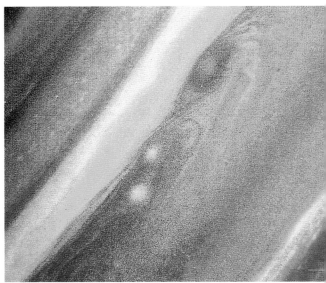

▲ Ribbon waves Exaggerated colour reveals ribbon-like structures within the lighter-coloured zones, and also swirling eddies that appear to transfer energy to the surrounding wind currents.

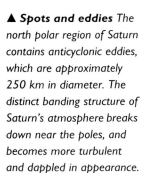

▲ Spots and eddies The north polar region of Saturn contains anticyclonic eddies, which are approximately 250 km in diameter. The distinct banding structure of Saturn's atmosphere breaks down near the poles, and becomes more turbulent and dappled in appearance.

▶ Filamentary clouds The bands at Saturn's mid-latitudes are often seen to interfere with each other. Shown here, an eddy of material which has broken off from a white band – in the shape of the numeral "6"– is ripped apart by a strong wind flow in the neighbouring orange band. The picture was obtained by Voyager 2 on 16 August 1981, from a distance of roughly 6 million km.

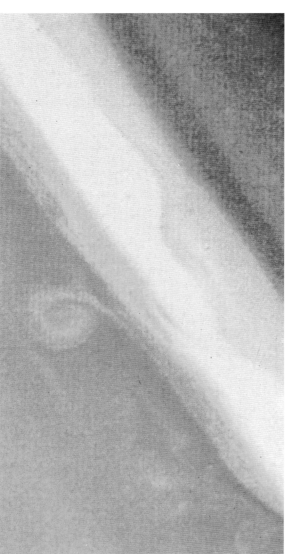

Storms on the horizon

Like Jupiter, Saturn is thought to possess a rocky core many times the mass of the Earth, surrounded by hydrogen at such a temperature and pressure that it behaves more like a conducting metal. Above that is a layer of liquid hydrogen, and a gaseous atmosphere whose main constituent is hydrogen, with some helium and traces of gases such as phosphine, ethane and acetylene. The upper atmosphere of Saturn is inherently less colourful than Jupiter's, although a high haze layer seems to cover up some of the cloud features lower down.

One puzzle, created by pre-Voyager infrared measurements, was that Saturn seemed to be generating more internal heat than expected. Because of its larger size, Jupiter has kept hold of more heat internally, but it is also shrinking, a process that generates additional heat. Saturn, being smaller, had less potential energy to begin with, and appears not to be shrinking. The answer seems to be that the hydrogen and helium in Saturn's inner layers are separating out, with the heavier helium descending under gravity. Infrared spectroscopy from the Voyagers revealed that there is only half as much helium in Saturn's upper atmosphere as there is in Jupiter's, and theoretical calculations confirm that this unlikely gravitational separation of hydrogen and helium may indeed generate enough energy to account for the relatively high heat output of Saturn.

Enhancement of the Voyager images eventually revealed many cloud features in Saturn's atmosphere. In fact, the weather changed perceptibly between Voyager 1's visit in November 1980 and

Voyager 2's in August 1981, with more storm systems present during the second encounter. Small eddies were seen, whose rotational energy "fed" the atmospheric circulation, giving rise to some of the fastest winds seen anywhere in the Solar System. Blowing parallel to the equator, these winds largely mimic the flows seen on Jupiter, although they were more evenly dispersed on either side of the equator. The winds were most ferocious at the equator, reaching speeds over 1,600 kilometres per hour. Material deeper down would rise to the surface, only to be disrupted by the winds higher up.

Sadly, there were no large-scale bursts of activity on Saturn similar to the large white spots seen occasionally from the Earth. Astronomers had to wait for a decade after the Voyager encounter for another major storm system, observed from the Hubble Space Telescope. Despite its faulty optics, the telescope's planetary camera took 100 images in November 1990, capturing the storm's genesis and rapid dispersal in the ferocious winds. It was the largest storm seen on Saturn in 60 years.

Much information was obtained when Voyager passed behind Saturn and its rings as seen from Earth. In particular, the fluctuations in the spacecraft's radio transmissions as the probe passed behind the rings and then the planet revealed interesting data. For controllers at JPL, this was the most nerve-racking time. With no contact while the craft was in the radio shadow of Saturn, it would not be possible to know if anything had gone wrong. Many fingers were crossed in the control rooms of the Deep Space Network (DSN).

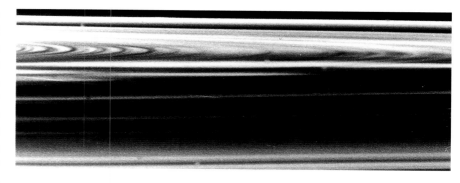

▲ **The rings side-on**
In an unusual view, taken from Voyager 2 on the day of closest encounter, the complexity of the ring structure was clear to see. The F Ring is at the bottom of the picture.

◄ **In the shadow of the rings** A marvellous view of Saturn and its rings, taken by Voyager 2 on August 11, 1981. The Cassini Division is clearly visible, as is the shadow cast by the rings on Saturn, showing just how opaque the rings are. A number of cloud bands are also present in this colour-enhanced view.

▶ **Back on target** In the early evening of Friday, 28 August 1981, Voyager 2 was able to return images of the planet Saturn after more than two agonizing days of transmitting nothing useful. The junk images (top) were caused by the jamming of Voyager's camera platform, followed by communications problems between JPL and the Australian Deep Space Network station at Tidinbilla. The first good image returned was a startling view of a crescent Saturn with backlit rings.

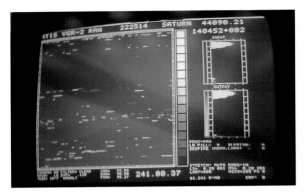

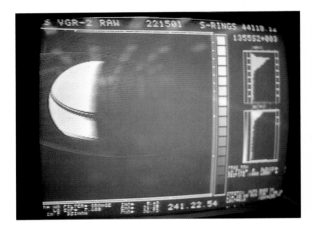

The first Voyager encounter passed without incident, unlike the second some nine months later. Just a few hours after Voyager 2's closest approach to Saturn on 25 August 1981, the mission's luck ran out. The spacecraft re-emerged from behind Saturn, only to return perfectly exposed images of empty space. In the anxious hours that followed, JPL engineers realized that the platform upon which Voyager's spectrometers and cameras were mounted had jammed.

There was speculation that the culprit was dust, which the spacecraft had met with during its passage through the rings; but the malfunction was later traced to a faulty component in the platform's gearing mechanism. The platform was jammed in such a way that it could move up and down, but could not scan across. Gingerly, DSN engineers attempted to free it, hindered greatly by the 160-minute delay for signals to travel from Earth to Saturn and back because of the limits imposed by

▶ **Next step forward** The joint NASA/ESA probe Cassini/Huygens should reach the Saturn system in 2004, after a seven-year journey using two gravity assists at Venus and one at Earth. NASA's Cassini orbiter will carry a suite of scientific instruments designed to make extensive observations of Saturn and its rings. After separating itself from Cassini, the smaller European Huygens probe will descend through the atmosphere of Titan to land on the surface.

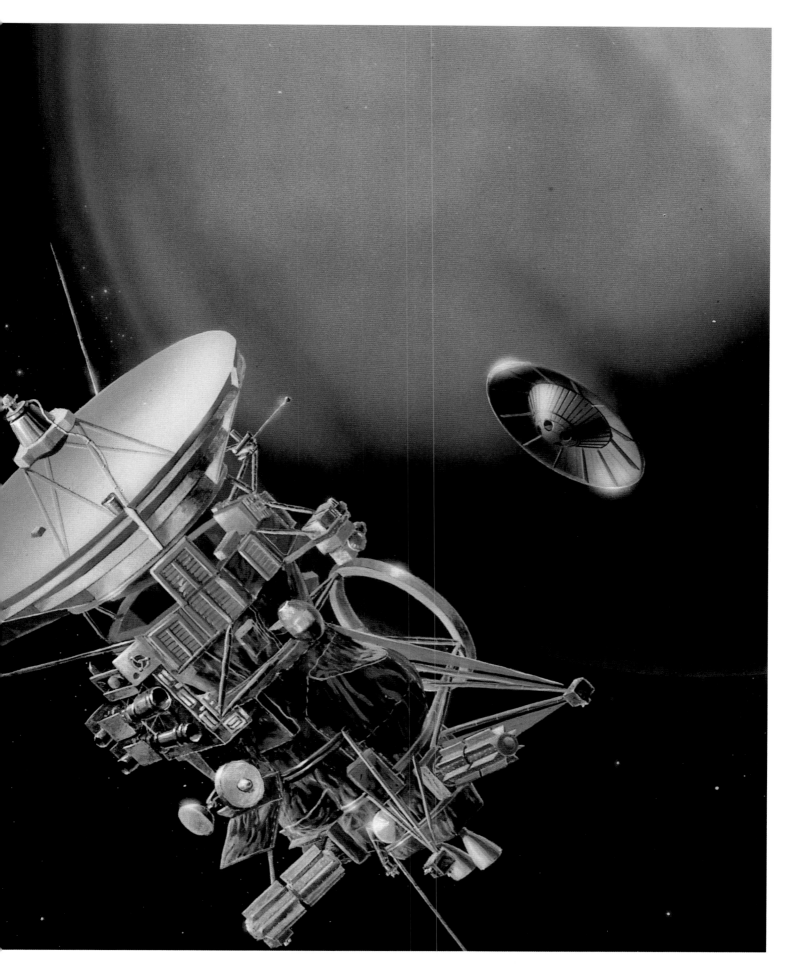

▲ Beneath the clouds
During the last few seconds of its parachute descent through Titan's dense atmosphere, the Huygens lander will see what promises to be one of the most exotic landscapes in the Solar System.

the speed of light. Finally, after two and a half days, they were successful. By the end of August 1981, Saturn was again in Voyager's view, this time a crescent with rings backlit by the Sun. Relieved engineers were to be found sporting T-shirts with the legend "Goodbye Saturn, Hello Uranus".

Now, however, Voyager 2 came up against a new problem in the mundane world of political reality. To a new administration in Washington, the

termination of the Voyager missions after the second Saturn encounter seemed the best way to balance NASA's budget. But to planetary astronomers it would have been the cruellest blow, because the spacecraft was now well on its way to Uranus and Neptune, with plenty of fuel and fully functioning equipment. In the end, however, intense lobbying saved the project, which now became known as the Extended Voyager Mission.

▲ **Backlit Titan** When Voyager 2 turned its camera towards a crescent Titan in August 1981, it found a halo of blue light extending around the moon. Several atmospheric layers seem to exist: An ultraviolet absorption layer at 400 km above the surface; a haze layer at 250 to 300 km; and a layer of smog-like aerosols at 200 km, which scatters blue light and obscures the denser atmosphere below.

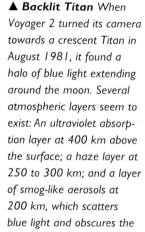

▲ **Gas-giant Titan?** A dark, circumpolar band is evident in Titan's dense atmosphere. The southern hemisphere also appears brighter than the northern.

▲ **Titan landfall** Initial studies had suggested Titan was covered with a massive ocean of liquid methane, but radar reflections suggest a mixed terrain of rock and icebergs floating in methane seas – the Huygens probe has been designed to survive landing on both solid and liquid surfaces.

▶ **Atmospheric haze** When Voyager 1 examined Titan, it saw an unbroken orange cloudscape above which was a blue haze. It was first assumed the atmosphere was methane, but Voyager revealed it be mostly of nitrogen mixed with hydrocarbons, including methane and ethane. Some of the methane may exist as separate cloud bodies.

The Cassini/Huygens mission

It is a measure of the way in which the fortunes of planetary exploration have fluctuated with the economic climate that more than two decades will have elapsed before the next spacecraft arrives at Saturn. Cassini/Huygens is due for launch, atop a Titan IV booster, in October 1997 and will not reach its target until June 2004, after two gravity assists at Venus and one at Earth. It will spend four years at the ringed planet, describing 63 complex orbits that will allow it to make close passes of all

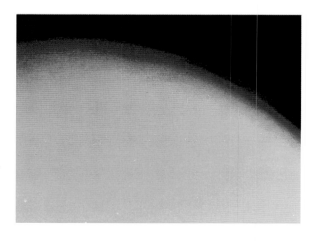

the sizeable moons, which it will scrutinize in more detail than ever before. The NASA-built Cassini orbiter will, in January 2005, dispatch the ESA-built Huygens probe to Titan, where it will descend through Titan's dense atmosphere in what promises to be one of the most spectacular events ever attempted in planetary exploration.

Titan is shrouded in a dense atmosphere, about 90 to 95 per cent of which is nitrogen, with up to 5 per cent methane, and traces of organic compounds like hydrogen cyanide. The presence of such compounds has led one observer to describe Titan as a "primordial Earth in deep freeze", because it is believed that our own atmosphere had a similar composition when life first emerged. Haze in Titan's atmosphere prevented much surface observation by the Voyagers, but intriguing results were gained. It was established that the surface temperature on Titan is around −180°C, and the surface pressure is roughly 1.3 times that at the Earth's surface. These conditions mean that methane can exist as a solid, liquid or gas. Huygens may descend into a breathtaking landscape where icebergs of methane float upon lakes of liquid methane. It will certainly be worth the wait.

KING GEORGE'S MAD WORLD

One pleasant spring evening in March 1781, an amateur astronomer in the quiet English country town of Bath discovered the seventh planet from the Sun and, in so doing, doubled the known dimensions of the Solar System overnight. But it was over two centuries after William Herschel happened upon the world, later named Uranus, before humans had any real understanding of the planet. Even by the standards of the outer Solar System, Uranus and its system of moons are weird. Today, about a decade after the Voyager 2 fly-by in January 1986, many aspects of the planet remain unexplained. It is probably no exaggeration to state (borrowing the

Crazed moon of Uranus *The undoubted highlight of the Voyager 2 encounter with Uranus was its bizarre moon Miranda. The contorted and jumbled surface of Miranda is a treasure-trove of geological oddities. A testimony to the cataclysmic events that punctuated the early history of the Solar System.*

words of Winston Churchill) that Uranus really is a riddle wrapped in a mystery inside an enigma. Such mysteries were hardly anticipated when the planet was discovered accidentally by Herschel using a 16-centimetre telescope he had built himself. Herschel had been an oboe player in the Hanoverian Army, but had settled in England and began to make his living by teaching music. By night he keenly scanned the heavens, and in March 1781 came across a faint disk that moved appreciably over the course of a few evenings. At first, Herschel believed he had discovered a new comet, but the final conclusion was inescapable: it was a planet.

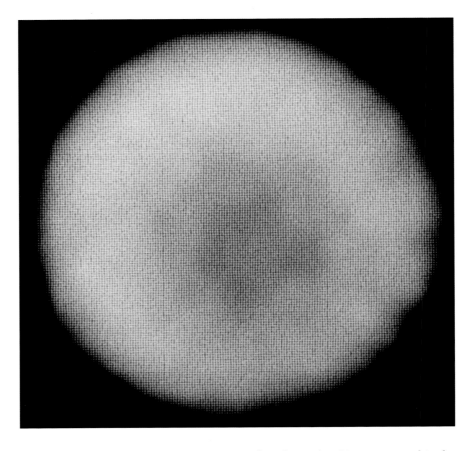

▶ Hardly the best
From Earth, some 3 billion kilometres distant, Uranus is little more than a blank disk. When the first CCD detectors came into use in 1976, the planet still showed no sign of surface features, even at infrared wavelengths which would reveal differences in surface temperature.

Herschel named his newly discovered world the "Georgian Planet", after the king of England, George III. But wiser heads prevailed, and it was eventually renamed Uranus, after the Greek god of the sky. Nevertheless, the king — who suffered repeated bouts of insanity — was delighted by his subject's celestial toadying, and appointed Herschel king's astronomer. Herschel himself went on to discover Oberon and Titania — two of the five moons known before the visit of Voyager 2. But, even with the best telescopes and the clearest viewing conditions, the planet remained a faint blue-green disk with virtually no visible features.

Uranus was found to take 84 years to orbit the Sun and to be roughly 20 times farther distant from it than Earth. Although Uranus was large, with four times the Earth's diameter, it paled in comparison to the gas giants Jupiter and Saturn. Observations of the moons orbiting the planet revealed the first of many mysteries: their paths were at right angles to Uranus's orbital motion around the Sun. Since the moons had to be orbiting close to the plane of the planet's equator, Uranus had to be tipped on its side. The axial tilt is actually 98 degrees, and so, during its progress around the Sun, first one pole and then the other faces the Sun directly.

▶ Rotation visible
When Voyager 2 began its observations of Uranus, very little surface detail could be made out, except by greatly overexposing the images. Eventually, thin clouds were seen at latitude 35° South. The motions of these clouds allowed the measurement of the atmospheric rotation rate. Uranus took just under 18 hours to make one rotation on its axis, which was then tilted towards the oncoming spacecraft.

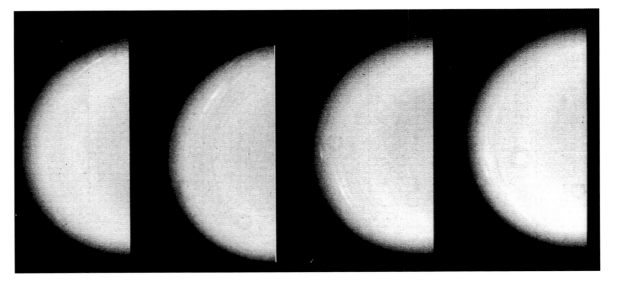

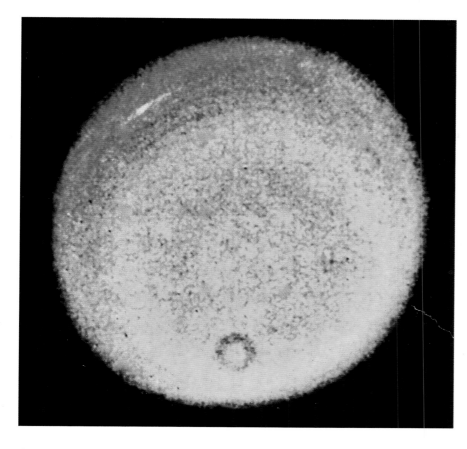

▲ Exaggerated colour
Extreme "stretching" of the Voyager images to bring out surface details reveals thin white elongated clouds on Uranus. Both clouds and planet move in a counter-clockwise direction. The ring-shaped blemishes are caused by dust on the camera optics.

Even with the advances in astronomical techniques during the Space Age, little new could be learned about Uranus. But, in March 1977, rings around the planet were discovered by astronomers high above the Pacific on board NASA's Kuiper Airborne Observatory, a converted transport aircraft. Uranus was observed as it passed in front of a star; but just before the limb of Uranus eclipsed it, the star's light blinked perceptibly five times, and then, when the "occultation" was over, the star dimmed a further five times. The only possible explanation was that there were at least five rings around Uranus, too faint to be seen with Earth-based telescopes. To say the very least, this was something astronomers wanted a closer look at.

So, before Voyager 2 was launched in August 1977, mission planners did their utmost to ensure that as much as possible would be learnt about Uranus if the probe survived to reach it. With the successful completion of the Saturn encounter in 1981, the Uranus encounter was on, and staff at JPL made the final adjustments to maximize the quality of data Voyager 2 would collect.

Voyager modifications

Uranus was twice Saturn's distance from the Sun, and receiving a quarter as much sunlight, so images of it would require longer exposures. But long exposures would lead to smeared images because of the speed at which the spacecraft was travelling. The planned close encounter with the moon Miranda, discovered as recently as 1948, would lose the most from image smearing, since Voyager 2 would pass just 29,000 kilometres above its surface. So JPL engineers devised a way of swivelling the whole spacecraft to compensate for its motion while the shutters of its TV cameras were open for up to 15 seconds.

The increasing distance of Voyager from Earth also reduced the rate at which data could be transmitted. So the data was "squeezed", which in the case of TV pictures meant transmitting differences in exposure between successive picture elements, rather than the exposures themselves. The back-up computer on board the spacecraft was instructed to do this automatically while the main computer was operating. On the ground, too, improvements were made: the main antennas of the Deep Space Network were electronically linked with adjacent dishes to pick up the faint signal from Voyager more efficiently. The project engineers and scientists hoped that these "fixes" would work for the

◄ True blue To the best approximation, this is how Uranus would appear to the naked eye. On 17 January 1986, Voyager 2 took this portrait of Uranus from 9.1 million km. Three frames were taken through orange, blue and green filters.

► Polar haze Extreme colour enhancement reveals a symmetrical hood of high altitude haze, extending from polar regions (centre) towards the equator.

◀ **Dusty rings** *Hitherto unsuspected "lanes" of fine dust were seen in this long exposure of the Uranian rings, made shortly after the closest approach by Voyager 2. Taken when the spacecraft was in the shadow of Uranus, the dust is seen to be spread throughout the ring system, and some of it is as bright as the rings themselves.*

close encounter with Uranus, which was scheduled for Friday, 24 January 1986. By the start of that year, it was clear that the ageing spacecraft – slightly deaf, slightly arthritic but still mentally agile – would be more than equal to the task at hand.

A topsy-turvy world

The first surprise of the Voyager encounter was how bland the atmosphere of Uranus appeared. To all intents and purposes, looking at Uranus was like gazing down into a bottomless ocean. Earth-based telescopes had revealed what appeared to be faint details of banding over many decades of observation, and Voyager scientists were expecting to see marking similar to those of Jupiter or Saturn. But from November 1985, when the spacecraft started surveying the planet in earnest, until the week of closest approach, just about the only thing visible on the planet was a slight darkening towards the poles. Voyager data suggests that the atmosphere is composed mainly of hydrogen and helium. Below this are thick, icy cloud layers of water, ammonia and methane. The outermost of these layers consists of methane, which absorbs red light and so gives Uranus its blue-green colour.

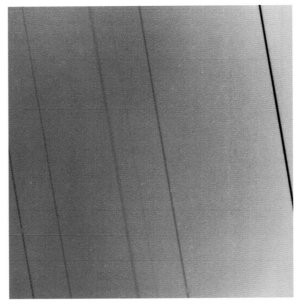

◀ **Darkness made visible** *The eleven rings of Uranus are blacker than soot, probably as the result of long exposure to the solar wind. They are difficult to observe at the best of times, and are seen most clearly against the backdrop of the planet, as in this Voyager 2 image.*

The lack of detail in the atmosphere persisted until the week before the close encounter. Then, by the most intense image enhancement, wispy clouds were found in the equatorial regions. But any relief at finding something was more than matched by a new riddle: the clouds were blowing the wrong

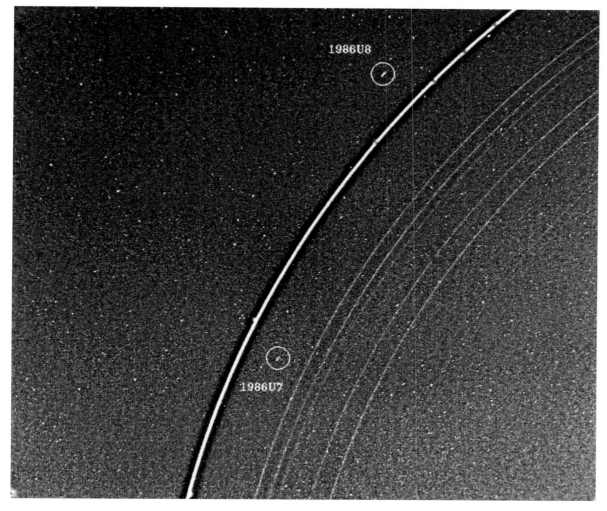

◄ Ring shepherds
Voyager 2 discovered ten new moons in and around the rings of Uranus; all have since been named after characters in the works of Shakespeare and Pope. Two of them, Cordelia (previously 1986U7) and Ophelia (1986U8) are shown here either side of the outermost ring.

▼ Psychedelic rings
"Star occultation" showed Uranus to have rings, but said little about how these rings would appear from close up, as indicated in this exaggerated view.

way. Before Voyager 2, meteorologists were certain that the unusual tilt of Uranus would give rise to peculiar weather and strong winds. It was expected that one of the poles would be the warmest place on the planet, since at the time of the encounter it was facing the Sun almost directly. Uranus's equatorial regions were almost in darkness whenever one of the poles was fully illuminated, and this was expected to create a strong thermal gradient that would generate wind patterns running in a north–south direction. But the thin clouds actually observed were moving parallel to the equator, as they tend to do on Earth.

Infrared measurements by Voyager 2 showed that there was, in fact, little temperature difference across the planet as a whole; the cloud tops were an unchanging −220°C. Clearly the planet had some unknown method of redistributing solar energy. Measurements of the total heat budget for the planet revealed that it radiated roughly the same amount of heat as it received from the Sun. So, with no appreciable internal heat source to drive a weather system, Uranus presents a dull face to the world.

In meteorological terms, the daily rotation of Uranus is more important than internal, or solar, heating. The clouds glimpsed by Voyager 2 revealed that the winds on Uranus blew at around 1,100 kilometres per hour, rather slower than on Jupiter or Saturn. By determining the internal rotation of the planet from magnetic measurements, it was possible to calculate that the Uranian "day" was slightly over 17 hours long, and so considerably less than the Jovian and Saturnian rotation periods.

The magnetic field of Uranus was perhaps the most peculiar aspect of a highly peculiar planet. Its strength was more or less as expected, but its orientation certainly was not, for the magnetic axis lies at 60 degrees to the axis

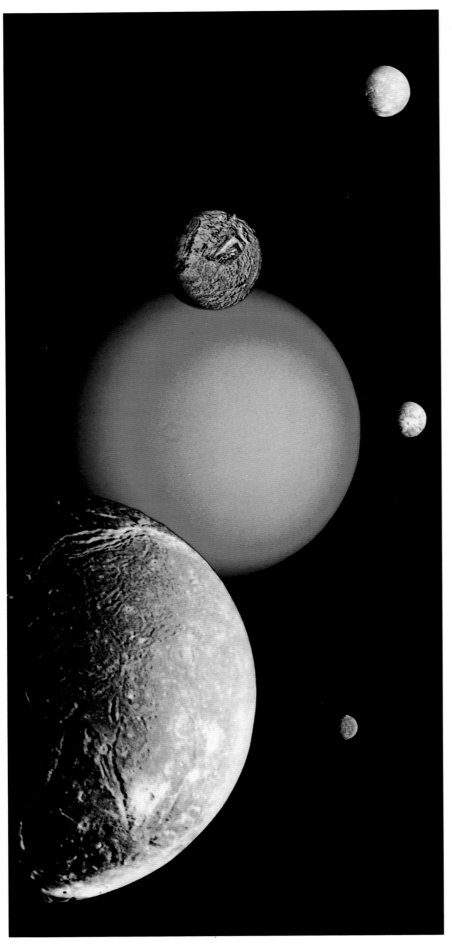

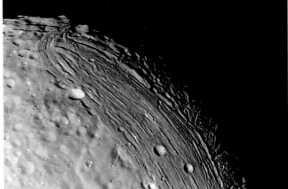

◄ Uranus and family A photomontage of the planet with its five largest moons: (from top) Titania, Miranda, Oberon, Umbriel and Ariel (in foreground). The images are not to scale: their diameters are 1,580 km, 470 km, 1,520 km, 1,170 km and 1,160 km. The moons are dark, heavily cratered, icy worlds.

▲ Grooves on Miranda The surface of Miranda consists of a strange mix of geological terrains. Lines of parallel ridges and grooves – termed "ovoids" – seem to be the scars of the repeated reassembly of the moon. There are relatively few craters, and those that do exist are relatively small and geologically young.

of rotation. It was like finding the Earth's north magnetic pole in Cairo, Egypt, and the southern pole in Brisbane, Australia. The magnetic field of Uranus is wound into the shape of a corkscrew by the planet's rotation. A hot, rocky core would not generate such a field, but surrounding the core is believed to be a "mushy" region, largely of water but with some ammonia, under such high pressure that the molecules break down into electrically charged ions. Currents set up in this electrically conducting "sea" could be responsible for generating Uranus's tilted magnetic field.

Dark rings and dark moons

That water ice predominates in this part of the Solar System is hardly surprising. Water ice seems to make up much of Uranus's rings and moons, as well as perhaps accounting for the planet's strange, axial rotation and magnetic field. After Uranus formed, there would have been many icy worlds the size of Earth orbiting nearby. If one such object hit Uranus, it could have knocked the planet sideways into its present oblique angle of tilt. Such an impact could also have led to Uranus losing more internal heat early on, which would explain why there is no strong internal heat source today. However, some astronomers believe that the interior is insulated in another, as yet unknown, manner.

During the Voyager encounter, the rings presented themselves to the spacecraft like the concentric

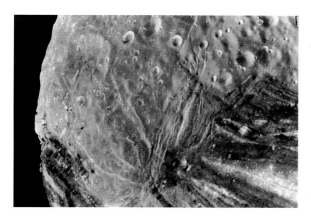

◀ Bizarre Miranda
Possibly the strangest moon so far found in the Solar System, Miranda was passed more closely by Voyager 2 than any other object on its journey, and details as small as 600 metres were seen on its weird surface. The curious jumbled terrain is well seen in this high-resolution image.

▶ The "Chevron"
A feature on Miranda that scientists have termed the "Chevron" is visible at left; it contains a light "V" shape, in the middle of an area of sharply defined, darker grooves. The Chevron probably has similar origins to the "ovoids" seen elsewhere on the chaotic surface of Miranda.

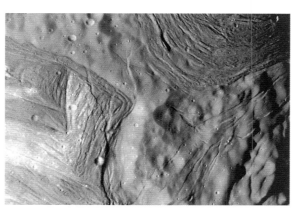

▶ Cliffs on Miranda
Escarpments catch the first rays of sunlight, revealing a cliff some 20 km high (more than ten times the depth of the Grand Canyon). Miranda seems to bear the scars of repeated shattering and reassembling in the period since the birth of the Solar System.

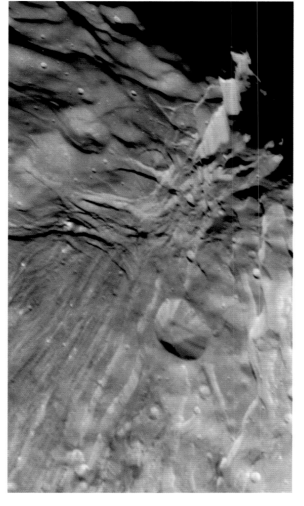

circles of a dartboard. From the Earth they were too faint to be seen directly, but Voyager 2 showed them to be very dark, absorbing some 97 per cent of sunlight. Only when backlit by the Sun could the otherwise pitch black rings be seen in any detail. Unlike the bright, youthful rings of Saturn, those around Uranus are probably quite ancient and have been darkened by prolonged exposure to solar radiation. Indeed, both of the Voyager craft found the solar wind to be remarkably blustery even at this distance from the Sun.

The moons of Uranus were also darker than their Saturnian counterparts. Voyager 2 found an additional ten moons within the orbit of Miranda, which until then had been the innermost of the five known moons. It is these small, dark inner moons that are the most likely source of material for the rings. Voyager found that Uranus is surrounded by eleven rings in total. The outermost of these, known as Epsilon, is slightly broader than the others and is distinctly oval-shaped, with two shepherd moons. A number of the remaining rings were also found to be oval in shape, and even slightly inclined to the planet's equator.

Because of the geometry of the Voyager 2 encounter, only the sunlit hemispheres of the five largest moons could be seen. Although all are less than half the diameter of our own Moon, they exhibit a wide range of geological features. The outermost, Oberon and Titania (the two discovered by Herschel), are the largest, and very similar in size. Titania has ice cliffs and fault lines, which hint at internal activity and an icy equivalent of tectonics. Oberon has signs of dark material in its craters, and what appears to be a 20-kilometre high mountain; but as this last feature was on the limb of the moon seen by Voyager, it may, in fact, be the central peak of a large impact basin.

The next two moons, Ariel and Umbriel, are smaller but no less interesting. Umbriel has a very dark surface with no signs of geological activity. Two bright spots were visible, presumably marking areas where underlying, brighter ice has been excavated by relatively recent impacts. Ariel is similar in size to Umbriel, but its surface is brighter with complicated patterns of valleys and grooves in which "lava" seems to have flowed. The lava cannot be molten rock, and is probably a mixture of water and ammonia extruded from the interior.

The jigsaw puzzle of Miranda
The highpoint of the Voyager's survey of the five larger moons of Uranus was undoubtedly Miranda. Although only 470 kilometres in diameter, Miranda looked, as imaging team scientists were later to declare, like a jigsaw puzzle of every other icy

◀ **Tragedy strikes** *On Tuesday 28 January 1986, Voyager scientists at JPL prepared for the last press conference of the Uranus encounter. On the other side of the USA, the space shuttle* Challenger *took to the skies on an unusually cold Florida morning. A second before the explosion that destroyed the shuttle, flames began to emerge from the joints of one of the shuttle's two solid-fuel rocket boosters.*

moon so far seen in the Voyager mission. By good fortune, Voyager 2 flew by close enough to take a detailed look at its unusual surface. Miranda was strewn with many new and incomprehensible features. Excited geologists had to invent new terms to describe them: "ovoids", "race-tracks" and even "layered cakes". The best explanation for this world – undoubtedly the riddle within the mystery of the enigmatic Uranian system – is that it was repeatedly broken apart by bombardment, and the fragments reassembled themselves. In Miranda, we may be seeing parts of the original core and surface jumbled together at random.

Miranda proved to be the answer to a long-standing mystery. The moons and asteroids in the Solar System, can be divided into two types: the larger ones that are nearly perfect "globes", and the smaller ones that are irregular in shape. With a diameter of 470 km, Miranda was on the boundary between the two types, and has large jagged features which could not persist for long on a larger body. Any smaller, and Miranda would perhaps have been an irregular potato shape; any bigger, and gravitational forces would have pulled it into a perfect sphere. The key was the process by which planetary material separates gravitationally into different layers – differentiation – giving a dense core and lighter mantle and crust (see page 29). The present diameter of Miranda is probably not quite big enough for differentiation to complete itself fully. In its earlier incarnations, the proto-Miranda may have been large enough to differentiate, and it is the resulting layers we see on its peculiar surface.

▲ **Challenger crew**
Seen on the morning of the ill-fated space shuttle flight, they are (from left to right): Ellison Onizuka, Christa McCauliffe, Mike Smith, Francis Scobee, Judith Resnick, Ronald McNair and Gregory Jarvis.

◀ **Frozen launchpad**
The day of the Challenger *launch was unusually cold, (one technician compared the launchpad to a scene from Dr Zhivago). The icy cold reduced the flexibility of the rubber O-ring seals on the rocket boosters, with disastrous consequences.*

▲ Farewell to Uranus
The crescent of Uranus provided a parting shot for Voyager 2. Uranus is normally featureless, but the whiteness of the limb suggested that there was a high-altitude haze above the main cloud decks. Meanwhile, scientists at JPL had already started to plan Voyager's encounter with Neptune, three-and-a-half years ahead.

The Challenger disaster

By any reckoning, the Voyager 2 encounter with Uranus was a triumph. Unfortunately, by the cruellest of ironies, it was overshadowed by a tragedy much nearer to home that changed the public perception of spaceflight forever. After a dearth of planetary missions, NASA had already declared 1986 to be the "year for space science". In one week in May, both the Galileo and Ulysses probes were to be launched on their respective paths towards Jupiter and the Sun's poles; and in September would come the long-awaited deployment of the Hubble Space Telescope. The launch of a schoolteacher aboard the space shuttle *Challenger*, at breakfast-time in California, reinforced the message that space travel was becoming routine.

For reporters and scientists at JPL, the spectacle of the *Challenger* launch was just a momentary distraction from the dramatic images streaming in from Uranus. A hiss of static and the announcement of a major malfunction shattered the myth of routine spaceflight. *Challenger* had exploded a minute after lift-off, killing the crew of seven. As a mark of respect, scientists at JPL postponed their final press conference, and some suggested naming features on the Uranian moons after the astronauts. In the end, they were commemorated on our own Moon.

The circumstances of the tragedy were investigated by a Presidential commission. The cause of the disaster was isolated to O-ring seals on the joints of the shuttle's two solid-fuel rocket boosters. Freezing temperatures on the day of launch had seriously reduced the elasticity of these O-rings, with catastrophic results. One of the commissioners, Nobel laureate Richard Feynman, suggested that a contributing factor was the unrealistically low assessment of the risks of space travel made by some NASA officials. He remarked, "For a successful technology, reality must take precedence over public relations, for nature cannot be fooled."

Blue World and Beyond

The golden age of planetary exploration came to a triumphant conclusion in August 1989 with the Voyager 2 fly-by of Neptune. Despite severe technical handicaps, the spacecraft had somehow survived the wear and tear of nearly 12 years of travel through space. Voyager 2's success at Uranus had given planetary astronomers a new-found sense of purpose, in stark contrast to the financial doldrums in which earlier Voyager encounters had taken place. If the early 1980s had been lean times for US planetary exploration, then the

The view from Triton
Within hours of the closest approach to Neptune in August 1989, scientists at JPL generated a view across the icy, geyser-spewing surface of its largest moon, Triton. Compared to Uranus, Neptune was far more meteorologically active, with distinct clouds in its deep blue atmosphere.

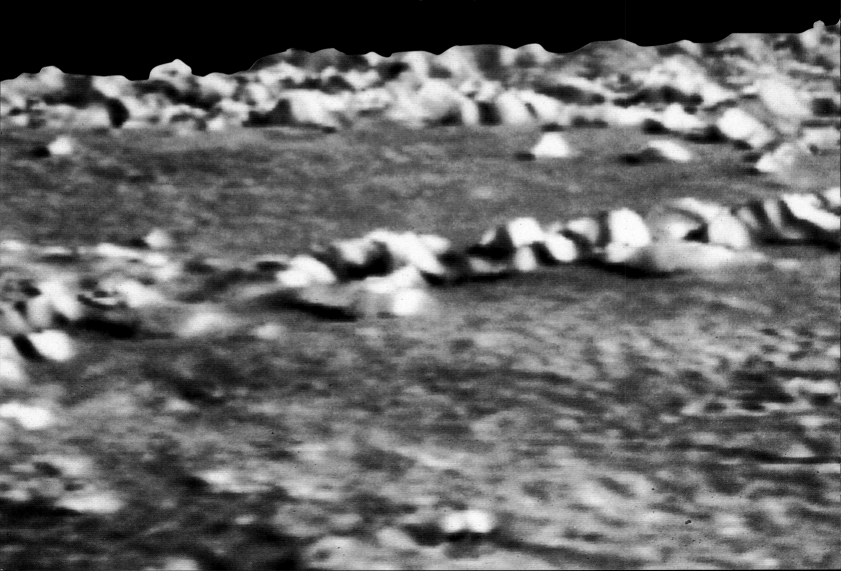

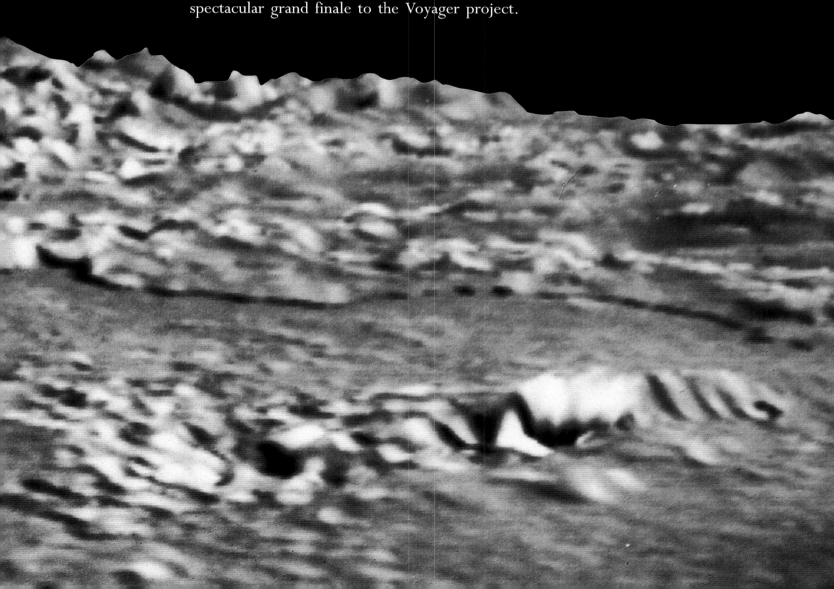

decade at least closed on an optimistic note: in 1989, after the two-and-a-half-year hiatus caused by the *Challenger* accident, things were looking up. By the time Voyager 2 reached Neptune, the Magellan spacecraft had been safely dispatched to Venus, and Galileo was ready to depart on its trek to Jupiter. The Neptune encounter itself presented a spectacular grand finale to the Voyager project.

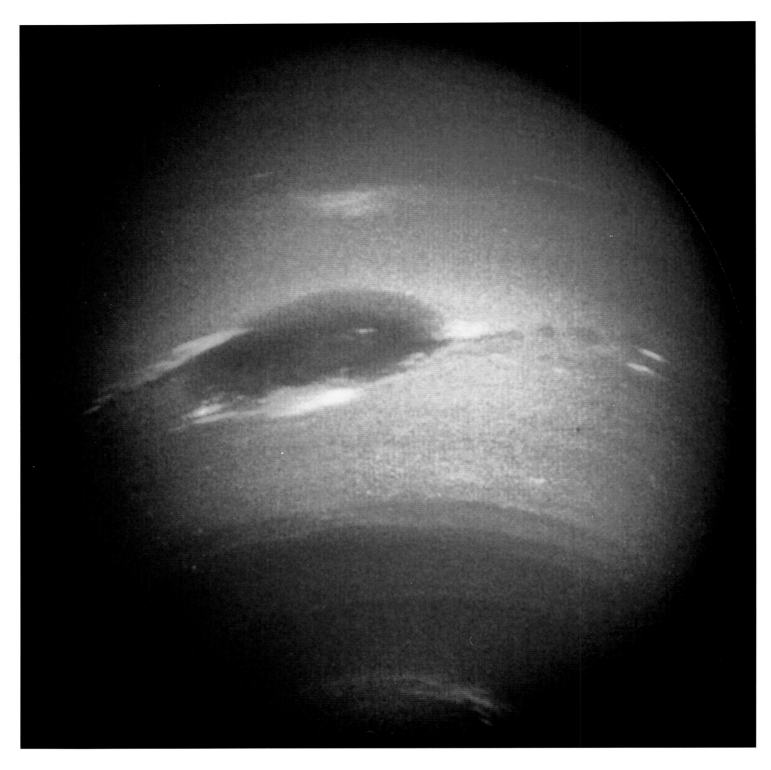

Voyager's final port of call turned out to be a remarkable place indeed. Neptune is half as far again from the Sun as Uranus, yet meteorologically it is far more active. Once again scientists gathered at JPL, for what many termed the last great hurrah of the Voyager missions. Neptune, it soon emerged, was surrounded by a peculiar ring system and was orbited by a moon on which geysers spouted nitrogen. The planet, named after the Roman god of the sea, was clearly putting on a show to close the final act of the Voyager project.

The long goodbye

By now it was taken for granted that the execution of a minutely timed encounter with Neptune would pass off flawlessly. After travelling nearly 5,000 million kilometres, Voyager 2 arrived only 10 kilometres from its planned rendezvous point high above Neptune's north pole on 25 August 1989. (In fact, Neptune turned out to be slightly closer to Earth than expected, and Voyager arrived 4 minutes 45 seconds early.) Many developments on the ground had made such precise interplanetary navigation

▲ **Great Dark Spot**
The most remarkable feature on Neptune was a storm system about the size of the Earth. Although Neptune is not as big as Uranus (its diameter is 600 km less), a stronger internal heat source is responsible for a far more active weather system.

possible, mostly the result of hard work by the engineers of the Deep Space Network (DSN). A spacecraft, which had left the balmy shores of Florida 12 years before, had survived the robotic equivalent of brain surgery to offset encroaching senility, and completed a course of speech therapy to make itself better heard from Earth.

At Neptune, illumination from the Sun is less than half that at Uranus. The technique, pioneered three and a half years earlier, of minutely choreographing long photographic exposures by slewing the whole spacecraft, would have to be further refined. From Neptune, Voyager 2's signals took slightly over four hours to reach home and were much weaker than from Uranus. By the time they reached the Earth, the signals were so attenuated that they were 20,000 million times less powerful than the output of a digital watch battery. The electronics within the Earth-based receiver stations had to be chilled to detect such faint whispers from the edge of the Solar System.

For the earlier Uranus encounter, dishes at DSN sites in California, Spain and Australia had been electronically linked together to "hear" Voyager's transmissions better. In order to receive signals from Neptune, DSN engineers had to rebuild many of their antennas, increasing their diameter from 64 metres to 70 metres. They also enlisted the help of the US National Radio Astronomy Observatory (NRAO) in New Mexico, to double the Goldstone tracking station's effective capacity to receive signals. By electronically linking the Goldstone dishes with the 27 separate dishes that form the NRAO's Very Large Array, it was possible to receive data from distant Neptune at roughly the same rate as it was received from Uranus.

The huge effort proved to be worthwhile, and Neptune did not disappoint the expectant audience worldwide. When the data began to come in, the eighth planet was transformed from a distant pinprick of light into a vivid globe, with an atmosphere far more active than that of Uranus. A few months before the August 1989 fly-by, Voyager's cameras were already making out cloud features. Neptune was blue, thanks to small amounts of methane in its hydrogen and helium atmosphere. A dark spot could be seen at roughly the same latitude as the Great Red Spot on Jupiter. It appeared to be the same size relative to Neptune as the Great Red Spot was to Jupiter, and could encompass the Earth quite easily. The feature became known, almost inevitably, as the Great Dark Spot. As Voyager 2 headed closer in, fine white clouds could be seen along its edges, reminiscent of cirrus clouds on Earth. Spectroscopic readings showed that the white clouds were composed of frozen methane crystals.

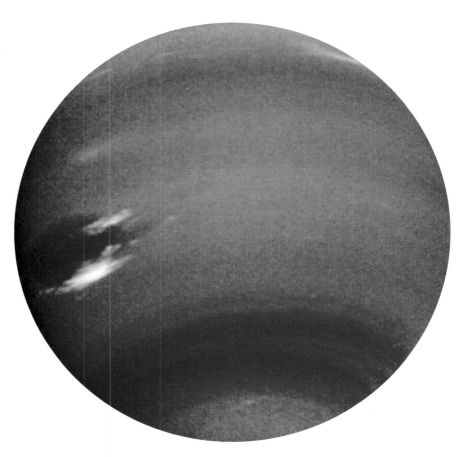

Riders on the storms

As Voyager 2 came closer in to Neptune, more spots could be seen in the atmosphere. A smaller dark spot closer to the south pole, christened D2, was seen, with white clouds descending into its centre. Intriguingly, D2 was rotating in a clockwise direction, whereas the motion of the Great Dark Spot was seen to be anticlockwise. Many of the

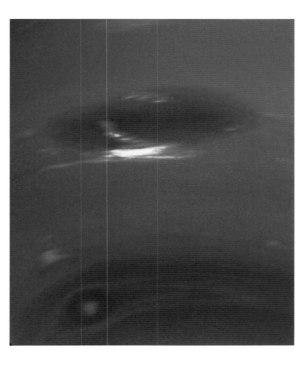

▲ Contrast enhanced
Absorption of red light by methane in the atmosphere gives Neptune its blue colour. Here, colours have been exaggerated to bring out features a few hundred kilometres across.

◄ Surface features
In addition to the Great Dark Spot, other notable atmospheric features seen on Neptune include the pale, triangular feature named the Scooter, and the small eye-like dark spot below it. Circulating above these features were bright clouds, probably of methane ice. The movements of such features have given important clues about the meteorology of Neptune.

▲ **Old hands** *Two of the more accomplished Voyager scientists show obvious delight at an image of Neptune on a computer* *at JPL. Bradford Smith, leader of the imaging team, points at the screen, while chief project scientist Edward Stone looks on.*

white clouds near the Great Dark Spot and elsewhere appeared to be at high altitudes because they cast shadows on the blue cloud decks below. Between the two dark spots another bright cloud feature was seen, which appeared to zoom around the planet at remarkable speed. It was nicknamed "the Scooter" – a little prematurely, it transpired, since it was later found to be almost stationary with respect to the rotation of the planet.

The big question was why is Neptune so much more dynamic than Uranus? The main reason seems to be that it generates more heat internally, and emits twice as much radiation at infrared wavelengths as it receives from the Sun. Indeed, the cloud tops of Neptune have the same temperature as those above Uranus, even though Uranus is much closer to the warming effects of the Sun.

The full answer was to be found deep in the Neptunian interior. The Voyager encounter enabled atmospheric scientists to piece together a picture of the workings of these inner regions. As befits its naming after the god of the sea, Neptune seems to consist mainly of a deep layer of water. By pure coincidence, when the planet was first discovered in 1846, one of the scientists involved in the search, James Challis, professor of astronomy at Cambridge, had suggested calling the new planet "Oceanus", an entirely apt description.

Beneath the deep layer of water lies a hot, dense core, probably of iron and rock. Though smaller in

▶ **Heart of darkness** *Feathery white clouds mark the boundary of the Great Dark Spot. Spiral features in the clouds and the* *boundary indicate that the spot is rotating in a counter-clockwise direction. Structures as small as 50 km across are visible in this image.*

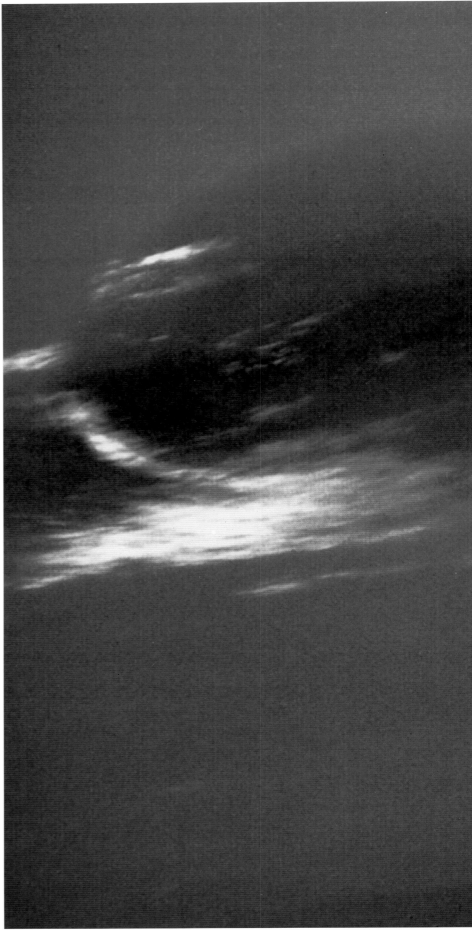

▶ South pole head-on

*A computer- generated
view of the south pole of
Neptune reveals a dark,
belt of cloud around the
pole, within which is a dark
spot at 55° South. This
spot took only 16 hours
to complete one rotation
— virtually the same as
Neptune's internal period
of rotation. At 42° South
is the Scooter, whose shape
changed from round to
square to triangular.*

▼ Casting shadows

*High altitude cirrus-type
clouds cast shadows on
the main deck of blue cloud
on Neptune some 50 km
below. The cloud streaks
are 50 to 200 km in width,
and many thousands of
kilometres in length.*

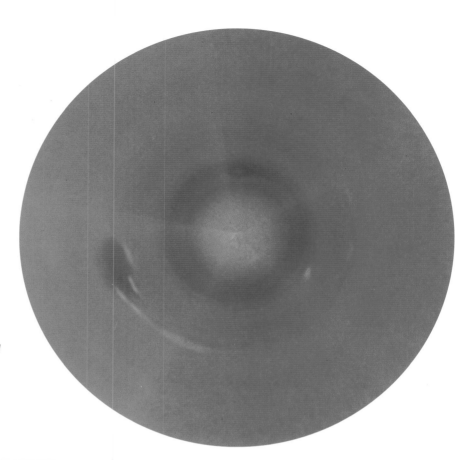

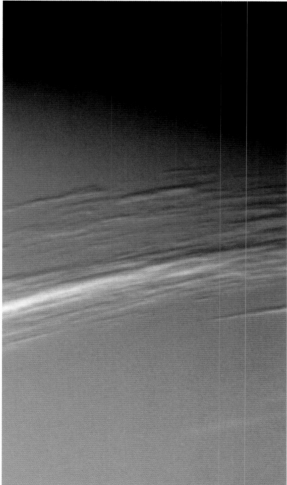

volume than Uranus, Neptune is more than 50 per cent denser, which suggests that the core is bigger. Calculations show that the core is probably hotter than the surface of the Sun, with a temperature of around 7,000°C. Two theories have been proposed to account for this heat, similar to those used to explain the conditions within Jupiter and Saturn. As with the inner planets, Neptune's heat could be derived mainly from the natural radioactive decay of rocks in the core. Alternatively, it could result from the separation of heavier elements and their descent towards the core — as occurs within Saturn. Further observations are needed to help decide which of the two possibilities is more likely.

Whatever their cause, the extreme pressures and temperatures within Neptune seem to break apart the chemical bonds deep within its watery oceans. Electric currents, generated in the resulting soup of charged particles, produce a powerful, yet curiously aligned magnetic field. The strongest currents are more than halfway out from the core, below Neptune's southern hemisphere. The net result is a magnetic pole located 47 degrees south of the equator. As it flew past the planet, Voyager 2 could sense the curious orientation of the magnetic field, which is very much like that of Uranus, although slightly weaker than had been expected.

As with Jupiter and Saturn, it is not possible to gauge the true length of the "day" on Neptune by observing its swirling cloud patterns. Regular radio

▲▼ Faint rings *The dark, grainy nature of the rings of Neptune is immediately apparent in these Voyager images. The two main rings are some 27,500 km and 38,000 km above the cloud tops, and a more tenuous, inner ring is 17,000 km out. A diffuse sheet of material seems to extend out from the central ring. All three Uranian rings contain tiny, micrometre-sized particles.*

emissions generated inside the planet by its magnetic field give the best figure for the internal rotation period: 16 hours and 7 minutes, a little shorter than the day on Uranus. When this was established, meteorologists were able to assess the true speeds of winds on Neptune. Astonishingly, the fast-moving feature known as the Scooter turned out to be static, in real terms, with respect to the rotation deep within the planet. The Great Dark Spot, it transpired, was zooming around in the opposite direction to the planet's rotation, at a speed of over 2,000 kilometres per hour – close to the speed of sound in Neptune's dense atmosphere. Quite how this speed is maintained remains a mystery.

The sausage-shaped rings of Neptune

Voyager 2's flight path took it to within 4,900 kilometres of Neptune's north pole, and then within 46,700 kilometres of its largest moon, Triton, which orbits Neptune backwards – that is, in the opposite direction to Neptune's rotation. Voyager's trajectory had been chosen to get as good a glimpse as possible of the whole of the Neptunian system, to determine just how typical a gas giant Neptune was. Did it possess a full set of rings, for example?

Before the encounter, there were two schools of thought on Neptune's ring system. The first was convinced that, because of Triton's unusual backward orbit its gravitational influence would disrupt any orbiting material before it could form into rings. The second group of scientists felt certain they would see incomplete rings, as Earth-based observations seemed to have shown. When Neptune had been observed passing in front of a star in 1984, the starlight was seen to dim perceptibly before the planet passed in front, but not after. The implication was that short arcs of material, rather than complete rings, were orbiting the planet.

It turned out that Neptune did have rings, but Voyager 2 was only able to see them in extremely long photographic exposures. Three weeks before the closest encounter, Voyager had observed what appeared to be ring arcs, and it was only a few days before closest approach that they were seen to be parts of complete rings. Curiously, they appeared more or less the same backlit by the Sun, as they did from the front. As one scientist later pointed out, the rings had half the reflectivity of soot and were seen against a black background, with under 900 times less illumination than falls upon the Earth. In all there were four extremely faint rings, the outermost of which was thicker and more "clumped" in places, appearing like sausages strung along a thin string. Small wonder that the rings had not been seen earlier in their entirety. It was later determined that these thickest portions in the outer ring coincided with the broad arcs of material observed from the Earth.

During its encounter, Voyager 2 made a quick dash through the plane of Neptune's rings. To the surprise of some scientists, as many as 300 impacts with the craft per second were recorded at 10,000 kilometres distance from the rings. There were impacts detected as far out as 50,000 kilometres. To the relief of all, Voyager escaped unscathed

In fact, Neptune's rings contain only about one-thousandth of the material in Uranus's faint ring system and, if assembled together in one place, they would form only a small body no more a few kilometres across. Although sparse, Neptune's rings are

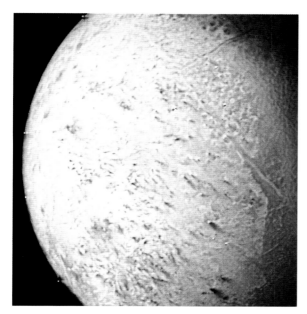

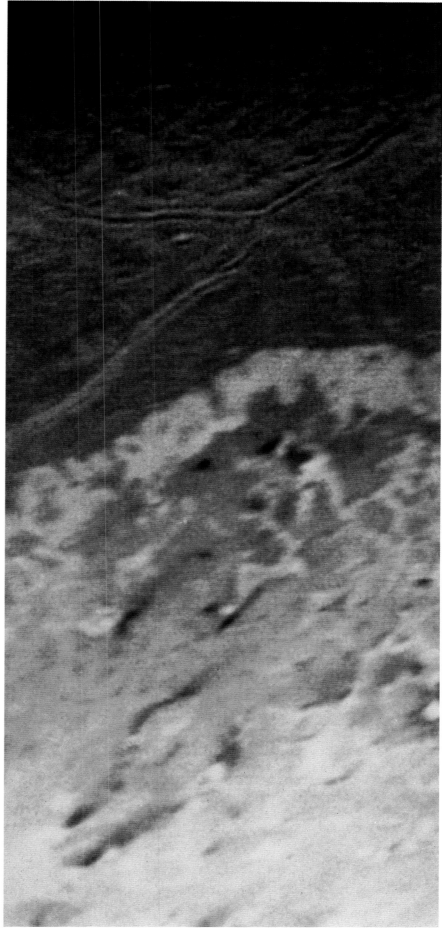

▲▶ **Triton** *The final port of call for Voyager 2 was Triton, Neptune's largest moon, which is some 2,700 km in diameter. A pink frost of methane covers the south polar ice cap. Green,* *violet and ultraviolet filters were used for this picture, giving the top area a false bluish tinge. The area away from the pole may consist of water-ice, which makes up the bulk of Triton's crust.*

composed of dark, icy objects which range from microscopic dust to the size of a small car. It is possible that even larger objects exist in the thickest "sausage" portions of the outermost ring, perhaps a few hundred metres across.

The existence of these thick arcs in the outermost ring was the source of a number of headaches. How were they formed, and why did they not even out? Voyager data suggested that roughly half the ring material is in the form of dust, ground down by the repeated collisions between the larger objects in the rings. These individual ring particles ought to slow down or speed up, depending on their size, with the overall result being to spread material more uniformly around the ring. However, the bulges seem to stay put. One suggestion was that the clumping is a recent phenomenon and may result from the impact of a comet upon a moon embedded within the rings.

The coldest moons

The Voyager encounter added six new moons to the two already known around Neptune. The most impressive of these, 1989N1, has since been named Proteus, and with a diameter of 420 kilometres was the largest moon discovered by Voyager 2. Proteus has one major feature, the Southern Hemisphere Depression, which is nearly 250 kilometres across and 15 deep and covers much of the hemisphere

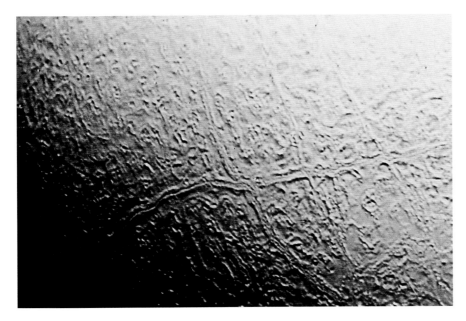

permanently facing Neptune. The other newly discovered moons are small, dark and icy, and in orbits close to the rings. It is thought that there may also be smaller shepherd moons closer to the rings themselves, but these have yet to been seen.

With a diameter of only 240 kilometres, the smaller of the two earlier-known moons, Nereid, was not seen well by Voyager 2. It circles Neptune in the most elongated orbit of any moon in the Solar System, taking almost a full Earth year to complete one orbit around its parent body. Even the best Voyager 2 image was not capable of resolving surface details. Thankfully, the same could not be said of Triton, Neptune's largest moon, which was to be the last great surprise of the Voyager mission.

Before Voyager 2, Triton had vied for the title of "largest moon in the Solar System". It was believed to be of huge size because it appeared to be rela-

▲ Cracks in the surface
Enormous cracks, roughly 35 km across, are found on the surface of Triton. They seem to be filled with fresh, "molten" ice, which has welled up from the interior in more recent times and then frozen. Triton's surface is the coldest place yet seen in the Solar System, with a temperature of −238°C, just 35 degrees above absolute zero.

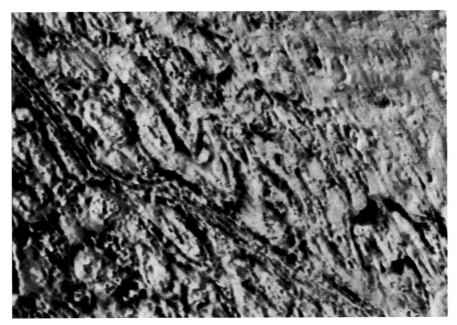

◄ Crazy cantaloupe
The textured and melon-like surface, christened cantaloupe terrain, is unique to Triton. Geologists believe that the closely spaced depressions and ridges may have resulted from repeated melting and freezing of the crust. The absence of impact craters on Triton suggests these processes occurred in recent geological history.

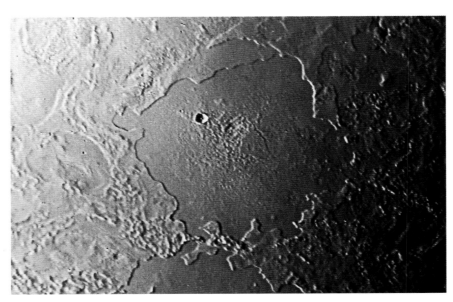

◄ Plain of ice What appears to be an ancient impact basin filled with ice is, in fact, probably the product of geyser-like eruptions of water and ammonia. The ledges at the edge of the basin may have been formed during previous activity. The opening pages of this chapter shows this feature in a perspective view.

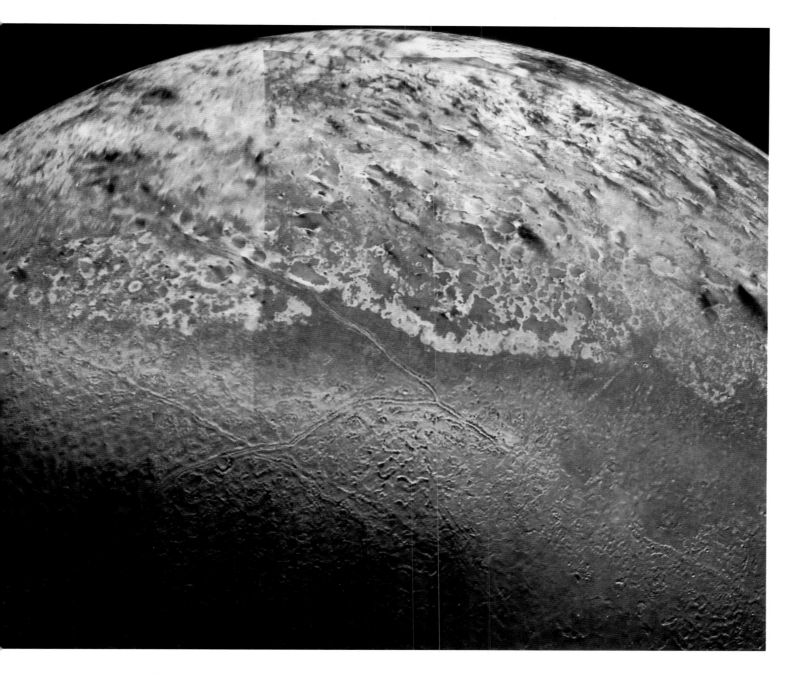

tively bright from Earth. But, after its discovery in 1846, very little could be gleaned about it because of its vast distance – apart from the fact that it orbits backwards. With improved spectroscopic techniques in the 1980s, Triton's surface was revealed to be icy, and composed mainly of methane and nitrogen; but its surface temperature could not be stated with any degree of accuracy, and astronomers were left to guess whether the nitrogen existed as a liquid, a solid or a gas. Some believed Triton would be similar to Saturn's moon, Titan, and would have a substantial atmosphere.

As Voyager 2 approached Neptune, it became possible to refine estimates of Triton's size. Disappointingly, the new figures showed that Triton was not, as long imagined, the largest moon of the Solar System, for its actual diameter turned out to be 2,700 kilometres (making it smaller than Earth's Moon, and way behind such giants as Ganymede and Titan). It also became clear that Triton was not covered by a dense atmosphere, but that its surface was exceptionally bright and icy – the cause of the earlier overestimate of its size. The atmosphere was very thin, composed of 99 per cent nitrogen with some methane, and with a pressure some 70,000 times lower than the Earth's. Infrared measurements revealed that the surface was frozen solid at a temperature around –238°C, and that some of the atmosphere had frozen out over Triton's polar regions. This thin frost has given these areas a bright pink hue, possibly the result of chemical changes in methane ice caused by cosmic rays. Triton was the coldest body seen by Voyager 2, and only more distant Pluto is assumed to be about as cold.

▲ **Pink icing** When Voyager 2 took this photo-montage of images of Triton, it was summer in the south polar regions. The pink colour of the polar cap may result from complex chemical changes to methane ice, produced by cosmic rays. The polar region is criss-crossed by darker streaks which result from "cryovolcanic" activity. The equatorial regions (at bottom) show a wealth of geological detail, highlighted on the previous page.

Triton's gassy geysers

The extreme cold made it all the more surprising when plumes of material were observed, spewing out several kilometres high into the tenuous atmosphere from "geysers" on Triton's surface. The erupted material was also seen to be diverted sideways by unexpectedly high winds. The presence of these geysers was revealed by dark streaks of "soot", strewn downwind like smudgy ink stains on a blotter. A month after the encounter, planetary geologists at the US Geological Survey (USGS) in Flagstaff concentrated their attention on the many dozens of these dark smudges. Voyager 2 had amassed a number of views of Triton from different perspectives, and stereo matching by computer revealed the progress and development of several geyser eruptions. The darkness of the streaks seems to result from blacker material carried aloft by the erupting nitrogen gas, and then floating down to the surface where it settles as a fine layer.

The energy source for this activity is all the more surprising, given its weakness – it is probably sunlight. Most of the surface of Triton is made up of a deep layer of ice, which acts much like the panes of a greenhouse. Heat passes through the ice layer, causing nitrogen below it to evaporate and accumulate. Eventually, the nitrogen pressure builds up until it is high enough to cause an eruption. The more pedantic geologists blanched at terming these eruptions "geysers", because that suggests the emission of liquid water. "Fumarole" is perhaps more correct, although it hardly conveys the spectacular nature of these jets of gas.

The rest of the surface of Triton has strange terrain types found nowhere else in the Solar System. There are flat plains resembling frozen lakes, fractures criss-crossing the surface for many hundreds of kilometres, and surface texturing which appears remarkably like the skin of a cantaloupe melon. How were these peculiar features formed?

The most plausible explanation is that Triton is geologically active, in the sense that the ice in the crust behaves, in some ways, like rock and lava. The curious landforms we now see on Triton's surface are the result of upwellings of molten material rising to the surface and then resolidifying. Scientists have even coined a term for this type of geological activity, "cryovolcanism", because it resembles volcanism, but takes place at the icy cold temperatures found in the outer Solar System.

What was the energy source for this activity? Only Jupiter's moons Io and Europa show similarly "fresh" surfaces, but this is caused by the continual gravitational tug-of-war between them. The energy to melt Triton's crust, it is widely believed, came from the satellite's capture by Neptune at some

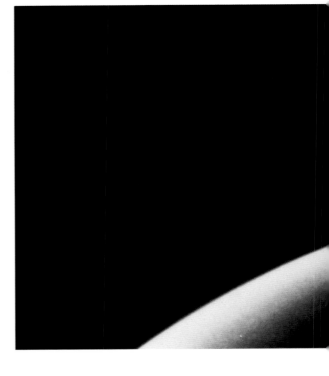

▲ **South polar storm**
After its closest approach to Neptune, Voyager 2 observed the south polar region as a thin crescent. A bright cloud feature was seen within about one and a half degrees of the south pole. This is thought to be part of the "eye" of a permanent storm system surrounding the pole.

◄ Goodbye Solar System *Perhaps the most poignant image of the whole Voyager mission was captured as Voyager 2 sped away from Neptune. The thin crescent of Triton rising above the crescent limb of its parent planet, providing a fitting farewell for a spacecraft now heading into interstellar space.*

time in the ancient past. This is partly corroborated by Triton's backward orbit, which would be hard to explain if it had formed along with Neptune in the same part of the Solar System. As Triton's orbit evolved in time to become more circular, the tidal forces generated would have been more than enough to melt the moon's surface layers.

Farewell and fare forward

Three days after its closest approach to the planet of the sea god, Voyager 2 returned a remarkable view of Neptune and Triton as two thin crescents. Of all the many thousands of images obtained during the Voyager mission, perhaps none is more evocative. It was the final shot before the credits rolled on one of the greatest scientific adventures of all time. For most of the Voyager scientists, the adventure was now over, and more than a few had tears in their eyes. "What a way to leave the Solar System," said Larry Soderblom, a geologist with the United States Geological Survey in Flagstaff.

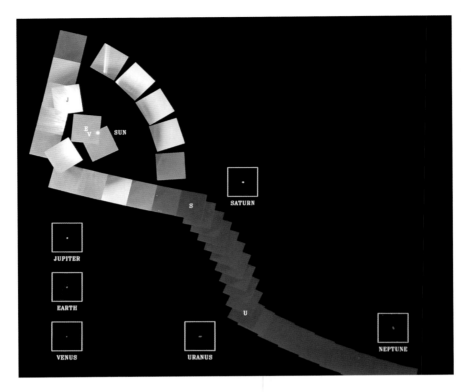

Since then, both the Voyager spacecraft have embarked upon what is euphemistically termed the "Voyager Interstellar Mission", with the ultimate goal of establishing the extent of the Sun's influence, and so determining the whereabouts of the "edge" of the Solar System. Even though many of their remote-sensing instruments have now been switched off, the Voyagers are continuously monitoring fluctuations in the solar wind with their radio receivers and plasma-wave instruments.

In the summer of 1992, the Voyagers began to detect curious radio signals. Scientists on the plasma-wave team calculated that their total power was 20 times the generating capacity of the United States. According to the team leader, Dr Don Gurnett of the University of Iowa, the signals emanate from the heliopause, where the solar wind slams into the cold interstellar gas. As the solar wind is not constant, the heliopause is irregular in shape, and it shifts with time. By tracing the solar wind at the time of the radio emissions, Gurnett and his colleagues estimated the location of the heliopause. Using the standard astronomical yardstick, the Astronomical Unit, or AU (the distance from Earth to Sun, 150 million kilometres) they put the heliopause 90 to 120 AU distant from the Sun. In early 1995, Voyager 1 was just 58 AU from the Sun, and Voyager 2 only 45 AU; so the probes still have a long way to go before they can measure the edge of the Solar System directly.

Of the four probes now on their way out of the Solar System, it might be thought that Pioneers 10 and 11 would be the first to cross the heliopause. They were launched four or five years before the Voyagers, and, having crossed the orbit of Neptune (Pioneer 10 in 1983, Pioneer 11 in 1990), have a head start. But the heliopause appears to have the shape of a ragged teardrop. Pioneer 10 is heading along the tail of the tear, while Pioneer 11 is travelling more slowly along its shorter axis. The problem is that power supplied to the craft by their onboard generators is decaying rapidly. Power margins for Pioneer 11 have already fallen close to the threshold below which it can no longer be "heard" from the ground, and officials at the NASA Ames Research Center estimate that nothing more will be heard from Pioneer 10 after the year 2000.

Power for the Voyagers does not become a problem until 2015, but by then they will be so distant that the Sun's dimness will be making it difficult for them to navigate with their Sun sensors. On the other hand, the ability of the Deep Space Network to receive ever-weakening signals improves each year and, with any luck, it may be possible to track the Voyagers until 2030.

Although we will eventually lose contact with these four messengers of human civilization, they will probably last more or less for ever, as long as they avoid colliding with large objects. They have so far withstood the solar wind and micrometeorite impacts; beyond the heliopause they will be at the mercy of high energy cosmic rays. But even such radiation is expected to cause minimal damage to the craft. In the foreseeable future, all four spacecraft will follow the motions of stars in our arm of the spiral Milky Way galaxy.

◄ **Destination unknown**
The Voyager spacecraft sets off on its interstellar journey in a computer-generated image by Pat Rawlings. JPL expects to track both of the Voyagers into the second decade of the 21st century, after which they will drift on forever, heading in different directions. Both carry the gold Interstellar Record as an extraterrestrial message in a bottle.

Out of sight and heading inexorably towards the stars, the longer-term legacy of the first four spacecraft to leave the Solar System is probably more philosophical than scientific. They all contain information about their origin. Pioneers 10 and 11 each have a small plaque, on which is inscribed the stylized trajectory of the craft relative to nearby pulsars and a picture of a man and a woman. The more sophisticated Voyagers contain similar locational information stored on gold-plated discs, along with images and sounds of Earth. If these exotic, interstellar equivalents of messages in a bottle are ever found by sentient beings, perhaps they will find them as thought-provoking and revealing as we found the myriad images the spacecraft returned to us in the first few years of their journeys.

Flotsam and Jetsam

Something odd has been happening at the edge of the Solar System. A crop of recent discoveries has effectively redrawn our perception of the outer Solar System and its farthest extremities. An enduring myth of planetary astronomy – the existence of a tenth planet, known as Planet X – has finally been laid to rest, and in its place has come the equally intriguing idea that the space beyond the orbit of Neptune is permeated by a multitude of small, icy planetesimals.

In 1990, analysis of the trajectories through space of the first pair of probes to exit the orbit of Neptune, Pioneers 10 and 11, revealed that their paths had not been affected by a tenth planet.

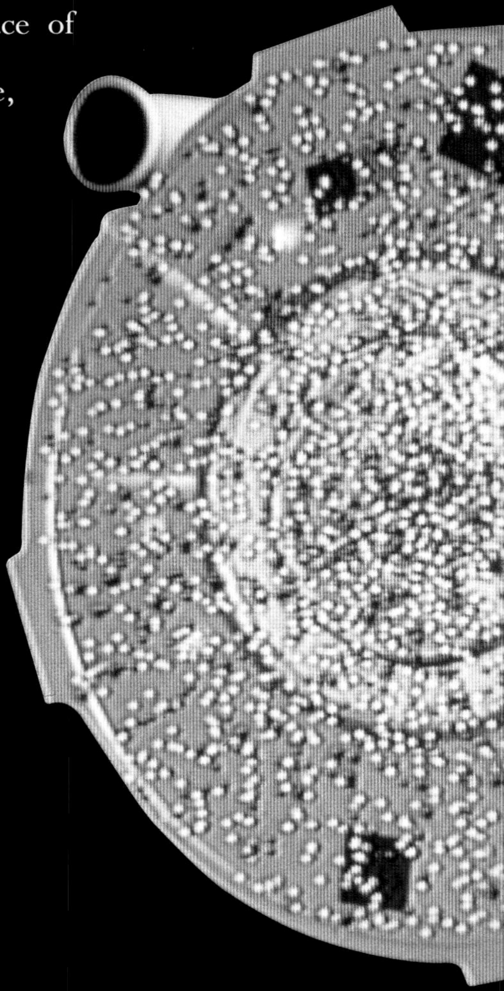

The survivor At about midnight GMT on 13–14 March 1986, the European Giotto probe passed within 600 km of the small nucleus of Halley's Comet, and survived a potentially devastating battering from dust particles. Shown here is a computer simulation of the damage done to the spacecraft's protective shielding. The probe's camera was knocked out but most other instruments survived, allowing Giotto to continue on to its secondary target, in order to make observations of Comet Grigg–Skjellerup in 1992.

More recently, calculations by astronomer Myles Standish have shown that the strongest evidence for the existence of Planet X – the wobble in the motions of Uranus – can be explained without the need for a tenth planet. After analysing the gravitational influence of Neptune on Voyager 2 during its fly-by in August 1989, Standish revealed, in the summer of 1993, that this influence is weaker than was previously believed. This discovery takes care of the discrepancy in the gravitational balance books of the outer Solar System. So, while this may not be the last word on the subject, it is fair to say that, to all intents and purposes, the idea of Planet X is dead.

Kuiper's children

In the place of Planet X has come the discovery of a number of small objects, suggesting that there is indeed a great deal of debris left over from the Solar System's birth in its outermost reaches. There are probably many thousands of small, icy bodies which collectively form a region known as the Kuiper Belt, named after the Dutch-American astronomer Gerard Kuiper. The first two were found as the result of a systematic search of inter-planetary space beyond Pluto, begun in 1987 at the University of Hawaii. Using the university's 2.2-metre telescope at Mauna Kea, astronomers have been scanning relatively wide fields of view at virtu-ally the limits of what is detectable, using the latest CCD detectors. At the end of August 1992, the British-born leader of the project, David Jewitt, and a colleague from the University of California at Berkeley, Jane Luu, found the first object in the Solar System ever detected beyond Pluto.

It did not seem like much: a small chunk of ice at most 240 kilometres across, and nearly 8,000 million kilometres from the Sun. The object was entered in astronomical catalogues as 1992 QB_1. With such an obscure, though entirely proper, des-ignation its significance was almost hidden, and it

▲ Halley's Comet

A bright comet in the night sky can be a dazzling sight. The nucleus remains hidden by the emission of gases and dust which form the comet's tail as it passes close to the Sun. This false-colour view of Halley's Comet was taken in April 1986 from Earth's southern hemisphere, from where the comet was best seen.

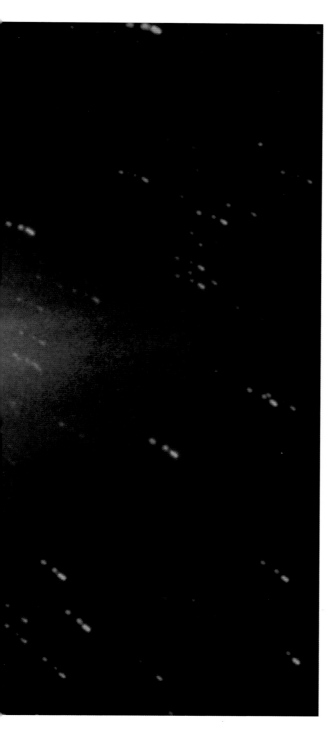

The mother of all snowball fights

Current theories suggest that the building blocks for the outer planets were millions upon millions of icy planetesimals, very large numbers of which aggregated together to form Uranus and Neptune. This process, however, was far from sedate and tranquil; in effect, during the earliest epochs of the Solar System, the region beyond Jupiter seems to have been the location of the mother of all snowball fights, on an interplanetary scale.

Over the course of many millions of years, countless numbers of planetesimals smashed into each other. Some survived to form larger objects, while others broke into smaller pieces. All were flung at random into intersecting orbits, and were either attracted or deflected by the immense gravity of the bodies that were to became the gas giants of the outer Solar System. Jupiter, in particular, seems to have catapulted large quantities of the icy material out of the Solar System, never to be seen again. Uranus and Neptune — much farther out and far less massive — flung many of these bodies into a vast shell of material surrounding the Solar System, anywhere between 50 and 100,000 Astronomical Units distant. Still bound by the Sun's gravitational influence, this possible extraplanetary resting place for icy debris is known as the Oort Cloud, after the Dutch astronomer Jan H. Oort who first proposed its existence, in 1951.

It is thought that icy bodies in the Oort Cloud are occasionally dislodged from their halo round the Sun when they pass through dense patches of interstellar gas and dust. They may also suffer gravitational perturbations from the movements of nearby stars in our arm of the Galaxy. As a result, these Oort Cloud bodies are either scattered outwards into interstellar space or inwards towards the Sun. Those that move inwards are accelerated towards the Sun and are captured into long, looping orbits. As they approach the Sun, their surfaces are heated and volatile gases and ice evaporate, taking dust particles with them, to leave a bright tail pointing away from the Sun. In this most familiar of guises, they are known as comets.

Comets have traditionally been classified according to their orbits. Some appear just once and head back into the outer Solar System, lost to view for ever. Others are captured into orbits with periods of 150 years or less, usually, it is thought, as the result of Jupiter's huge gravitational influence.

This standard picture of cometary origins has, however, recently been called into question. The idea of the Oort Cloud is by no means universally popular, and some astronomers dispute its existence. A major problem is that short-period comets should enter the Solar System from all directions.

only hit the headlines when Luu and Jewitt christened it less formally — as devotees of the novels of John Le Carré, they named their discovery "Smiley". The name was perhaps appropriate, for their search had been as labyrinthine as one of Le Carré's plots. In March 1993, they found another, similar object and called that Karla. "The excitement now is that we can hope to study the building blocks of the outer planets" said co-discoverer David Jewitt, "These are the most primitive objects we are likely to find in the Solar System." At the time of writing, about two dozen such Kuiper Belt objects were known, and more are being found at a remarkable rate.

► The heart of Halley
Surface features on the
nucleus of Halley's Comet
were revealed by ESA's
Giotto spacecraft. At left, a
"crater" is visible between
the brightest emission
regions, as is a sunlit
"mountain" towards the
night side. The picture is a
composite of seven images
taken by the Halley Multi-
colour Camera on the night
of 13–14 March 1986.

If the Oort Cloud exists, it should be prone to gravitational disturbances from all directions and not in certain ones preference to others. Yet most short-period comets are in orbits close to the ecliptic (the plane of the planets) and travel in roughly the same direction. This coincidence seems to contradict the idea of random disturbances in a distant Oort Cloud. The Kuiper Belt is now more favoured as the region in which short-period comets originate, adding a further twist to the already complicated picture of the outer Solar System.

Rock and ice

There is also less certainty about another equally standard theory connected with the birth of the Solar System. The detritus left over from the gene-sis of the planets has traditionally been separated into two distinct sorts. The predominantly rocky material was believed to inhabit the asteroid belt, while the mainly icy material was thought to reside in the outer Solar System and occasionally make a fleeting visit to the inner regions in the form of comets. This view has now changed, thanks particu-larly to the results of space missions to perhaps the

most famous comet of all, Halley, which sped through the inner Solar System in early 1986.

To greet this regular celestial visitor, a veritable flotilla of spacecraft was sent up. Two were Japanese, the Suisei and Sakigake probes; two were Soviet, Vegas 1 and 2; and the last, and by far the most sophisticated, was the European Giotto probe. The missions were undertaken in a new spirit of international co-operation. Using data gathered dur-ing the earlier fly-bys of the other craft to enable accurate navigation, Giotto was able to pass within 600 kilometres of Halley's nucleus at midnight GMT on 13–14 March 1986. The nucleus turned out to be a cratered peanut-shaped body just 16 by 8 by 8 kilometres in extent.

Most surprisingly, Halley is extremely dark, absorbing some 96 per cent of the sunlight falling on it, making it one of the darkest objects in the Solar System. Heat from the Sun cracks open its dark crust to reveal fresher material within, and bright jets of dust and gas issue from the cracks. Intriguingly, Giotto showed that this ejected materi-al ejected contains carbon, hydrogen, oxygen and nitrogen – the building blocks of life – in various

◀ **Giotto sequence**

Six images taken as Giotto moved closer to the nucleus of Comet Halley from 25,600 km to 3,900 km. The dark nucleus is visible against the brighter jets of material streaming from the surface at two main sites. The nucleus measures about 16 km by 8 km by 8 km, yet the comet's "tail" of evaporated material can extend out for hundreds of thousands of kilometres.

chemical combinations. The emerging consensus is that comets are not dirty snowballs, as originally assumed, but icy dirt balls.

It has been estimated that Halley's Comet loses nearly 250 million tonnes of material at each return to the inner Solar System. At this rate, the comet will be totally depleted of its ices and gases in 200,000 years, leaving behind a husk of rocky material. All short-period comets suffer a similar periodic heating and buffeting by the Sun, and their volatile material is progressively depleted. There is now a growing belief among astronomers that many asteroids — particularly those that cross inside the Earth's orbit — are actually "dead" comets that have expended all of their volatile components. As we shall see in the next chapter, this intriguing possibil-

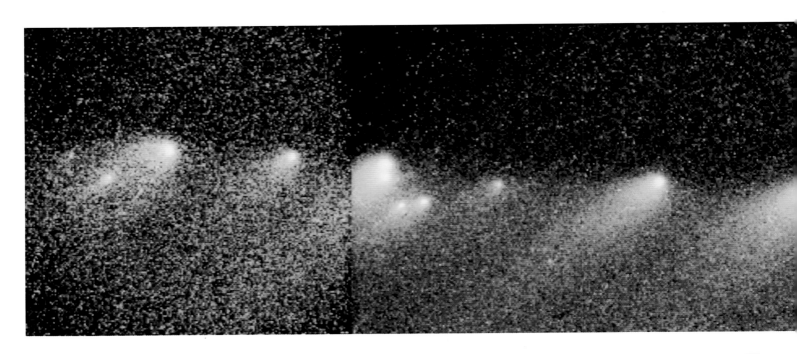

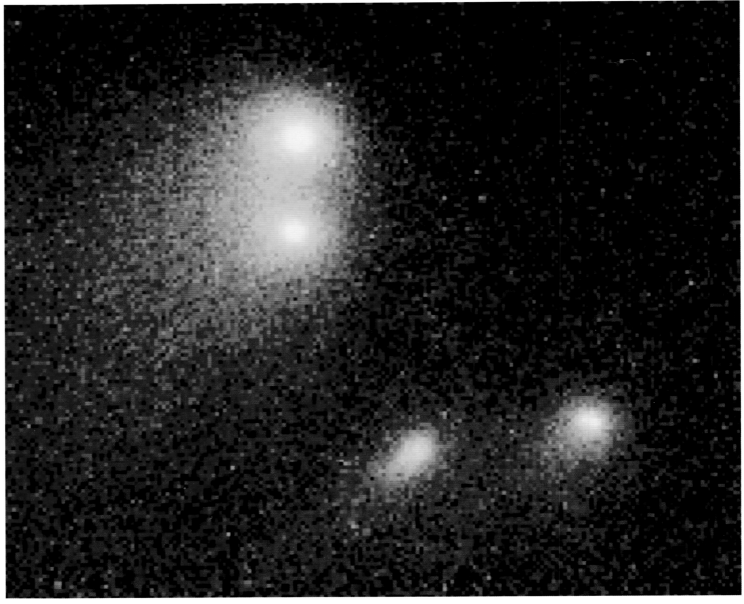

◀▲ *Shoemaker-Levy 9*
Billed as the astronomical event of the 1990s, Comet Shoemaker-Levy 9 collided with the giant planet Jupiter in the third week of July 1994. As can be seen (at top) in a composite image from the repaired Hubble Space Telescope, the comet had been disrupted into more than 20 separate fragments by the powerful gravitational influence of Jupiter. At left, is a more detailed view of the brightest fragments of the comet, taken with Hubble's newly serviced Wide-Field Planetary Camera in January 1994.

ity may be addressed in one of the first missions in NASA's low-cost Discovery programme, which was planned for launch in 1996.

The Giotto craft survived its Halley encounter to make a close pass of an older, less active comet known as Grigg–Skjellerup, in July 1992. The European Space Agency (ESA) is now in the process of identifying a number of possible cometary targets for study on a mission to be launched in 2003 or 2004. Known as Rosetta, the mission will involve a landing on a cometary nucleus to take the first samples of cometary material. As its name suggests, the Rosetta mission will hopefully be able to decipher the mysteries of the Solar System's history from the evidence of an ancient interplanetary artefact.

Pluto and Charon

In the context of the most pristine material in the Solar System, it is now time to introduce Pluto and Charon. Since its discovery, Pluto has exasperated planetary astronomers because of its peculiar characteristics. Pluto is the smallest, coldest and quite possibly the weirdest of all worlds. It is not always the farthest planet from the Sun, for its orbit is extremely elongated and it varies between 4,400 and 7,400 million kilometres from the Sun bringing it inside even the orbit of Neptune. The orbit is also inclined at 17 degrees to the plane of the other planets. Until the 1970s, very little else was known about Pluto until sensitive infrared spectroscopes discovered methane on its surface.

In June 1978, James Christy of the US Naval Observatory discovered a moon roughly half the size of Pluto, orbiting very close to the planet. Planet and moon together perform a subtle gravitational waltz around a common centre of gravity.

Like the Earth and Moon, their orbits are "locked" relative to each other with a period of 6 days 9 hours. Charon's discovery was extremely timely, for the plane of its orbit soon passed through the line of sight of Pluto as seen from Earth. The result was that, during the years 1985 to 1990, Pluto and Charon eclipsed each other regularly, leading to a wealth of new discoveries, all the more remarkable as this mutual eclipsing is seen from Earth only twice in Pluto's 248 year-long orbit of the Sun.

The short spell of mutual eclipsing allowed astronomers to watch Charon pass in front of and then behind Pluto. By measuring the dip in light as they eclipsed each other, it was possible to determine accurately their respective diameters. The total mass of the system was calculated from their period of rotation and, since their diameters were now known, it was possible to get a good estimate of their densities, and therefore compositions.

It is now clear from the spectral analysis of light reflected from Pluto and Charon that the objects are different. Charon is covered by greater amounts of water ice, while Pluto's surface is covered by the exotic ices of nitrogen, carbon monoxide and methane. As they are two of the coldest bodies in the Solar System (at around –225°C), it is possible to guess what their surfaces would look like from the strange physical properties of ices at this extreme temperature. The water ice on Charon would be as rigid as rock, so there could be escarpments and cliffs of solid ice. The more exotic ices on Pluto would be structurally weaker and perhaps slushier, with no distinct landforms. Charon has fewer exotic frosts because of its smaller size: its weaker gravity cannot hold on to them, and they have long since evaporated away into space.

◄ **Twin worlds** *From Pluto and Charon, seen in this artist's impression, the Sun is just a bright star. The two worlds are locked into a synchronous orbit around each other, taking 6.4 days to complete one full cycle. Charon is roughly half the diameter of Pluto, and is the largest moon in proportion to its parent body in the Solar System.*

▼ **Hubble views** *The unrepaired Hubble Space Telescope in 1990 was able to resolve Pluto and Charon with greater clarity (right) than the ground-based Canada–France–Hawaii telescope (left). At the time, Charon was near its maximum orbital separation as seen from Earth.*

Pluto is decidedly brighter than Charon, and its poles seem to be more reflective than other parts of its surface. As most planetesimals are dark, some astronomers believe that Pluto's fresh appearance is linked to the presence of an atmosphere. In June 1988, observations from NASA's Kuiper Airborne Observatory (from which the rings of Uranus had been detected a decade earlier) revealed a gradual change in the brightness of a star before and after Pluto passed in front of it. If Pluto were completely devoid of an atmosphere, the star's light would have switched off and on again very abruptly. This suggested that Pluto has a tenuous atmosphere, with a pressure perhaps only a few millionths of that at the surface of the Earth.

Although from Pluto the Sun is so distant that it would appear only as a relatively bright star, it still provides enough heat to evaporate some of the surface frost, creating a transitory atmosphere. The atmosphere is at its most extensive when, as now, Pluto is closest to the Sun. From the 1988 observations, this atmosphere appears to consist of nitrogen and methane and may extend out far enough out to envelop the moon Charon.

More recent observations of the Pluto–Charon system were made in 1990 by the Hubble Space Telescope. The latest estimates for their diameters are 2,284 kilometres for Pluto and 1,192 kilometres for Charon. These figures indicate that Pluto is twice as dense as water ice, so there must be a

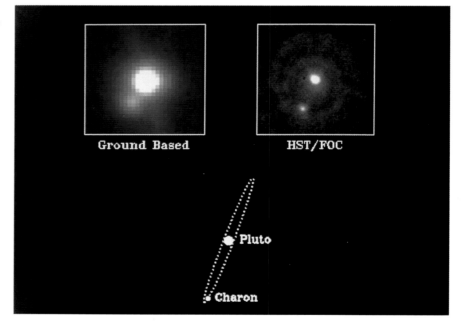

good proportion of rock within it. This could perhaps have been predicted after the Voyager 2 pass of Triton, which has similar proportions. We are left to guess whether Pluto is in other ways similar to Triton — whether its outer layers have melted completely, or if its surface is strewn with active geysers spewing frozen gases high into its tenuous atmosphere. As we shall see in the last chapter, we can hope that it will be only a few years before we get our first close-up glimpse of these strange worlds.

► **Mission to Pluto** *By the end of the last decade of the 20th century, JPL engineers may have built the Pluto Fast Fly-by spacecraft. With a total "fuelled-up" weight of 162 kilograms, the craft should reach the Pluto–Charon system by the year 2010.*

THE JOURNEY CONTINUES

"Houston, *Endeavour*. We have a firm handshake with Mr Hubble's telescope."

With these words from astronaut Dick Covey, one of the more unusual episodes of space exploration in recent years began. The date was Saturday, 4 December 1993, and the location 595 kilometres above the Earth, where the space shuttle *Endeavour* had successfully made its rendezvous with the Hubble Space Telescope. Over the next week, the shuttle crew had the important task of "servicing" the telescope, which was suffering from greater problems than just the wear and tear of three years in space. For besides repairing the telescope, the unspoken purpose of the *Endeavour's* mission was to restore public confidence in the ailing US space programme. The embarrassing revelation that the telescope had been launched with faulty optics continued to haunt NASA, despite the valuable scientific results obtained by astronomers who had learnt to work around the optical problem with advanced image processing techniques.

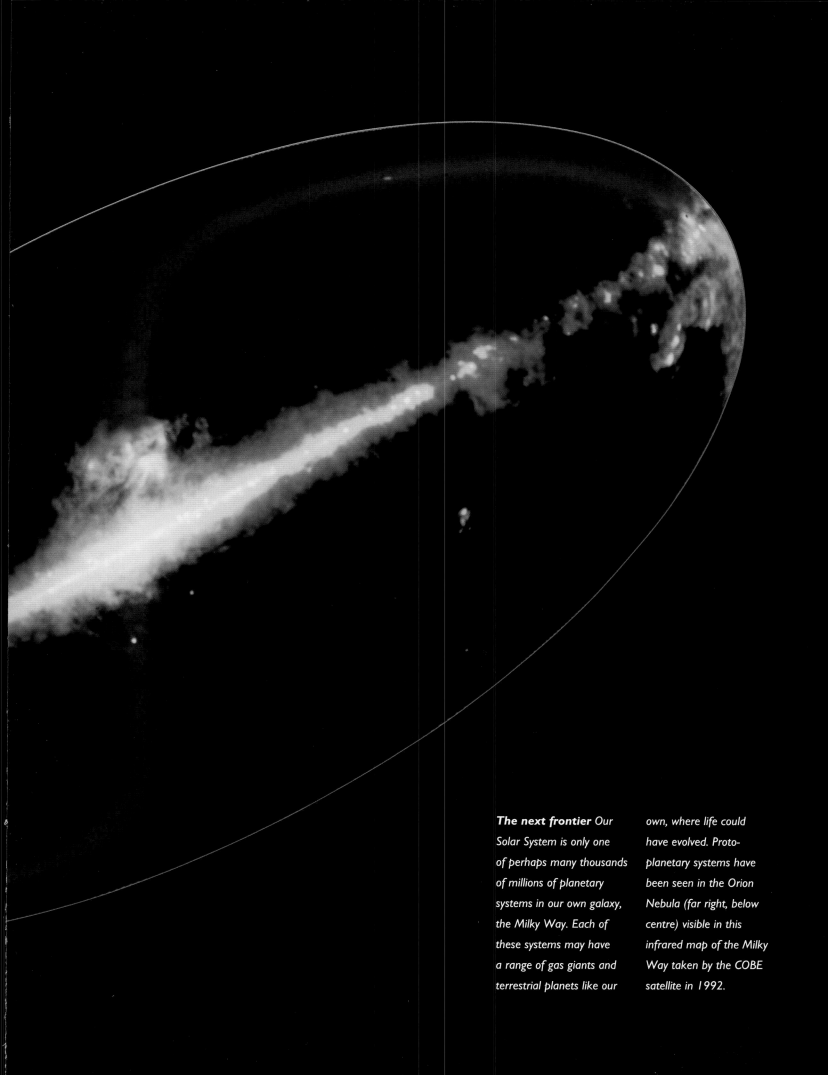

The next frontier Our Solar System is only one of perhaps many thousands of millions of planetary systems in our own galaxy, the Milky Way. Each of these systems may have a range of gas giants and terrestrial planets like our own, where life could have evolved. Proto-planetary systems have been seen in the Orion Nebula (far right, below centre) visible in this infrared map of the Milky Way taken by the COBE satellite in 1992.

◄ **Hubble repair mission** *On 2 December 1993, at 4.27a.m. local time, the space shuttle* Endeavour *took off from Pad 39B at Cape Canaveral, Florida. The seven astronauts on board formed one of the most highly trained shuttle crews ever; their task was to refurbish the Hubble Space Telescope in orbit.*

► **Hubble pit-stop** *On Saturday, 4 December 1993,* Endeavour *caught up with Hubble in orbit and, seen against the backdrop of Madagascar, successfully brought the telescope into its large payload bay.*

▼ **At work** *Over about a week, astronauts performed many hours of work on the Hubble Space Telescope, fitting two new optical instruments and servicing the remainder of the telescope. Seen here is shuttle astronaut Kathryn Thornton on the fourth day of extra-vehicular activities. Alternating pairs of space-suited astronauts made extended spacewalks over a period of five days.*

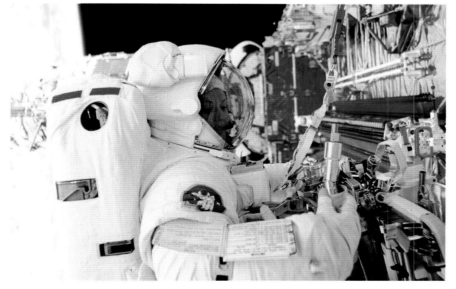

In short, Commander Covey's crew aboard shuttle mission STS-61 had the daunting task of bringing the Hubble Space Telescope's blurred images into focus. Their task read like something from the 1970s TV series *Mission Impossible*: replace Hubble's twin solar panels, which had caused the entire telescope to jitter; perform a series of delicate "housekeeping" operations around the telescope; and finally, remove two large, piano-sized instruments and replace them with new ones, to correct the telescope's well-known spherical aberration which splayed light rather than bringing it to a crisp focus. One of the replacement instruments, known as the Wide Field Planetary Camera II or "Wifpic 2", had been built at JPL. Its optics were designed to compensate for the spherical aberration. The other new instrument, known as the COSTAR (Corrective

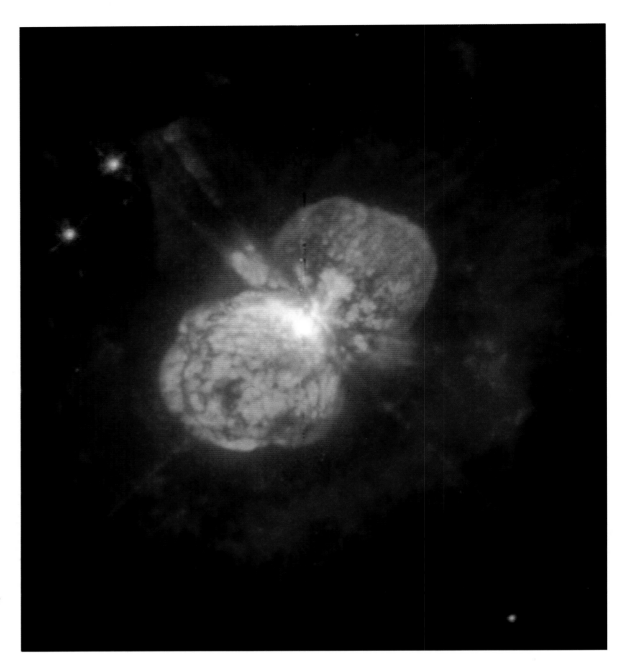

▶▶ **Hubble pay-off I**
One of the beneficial results of the repairs to Hubble came from the improved optics. Before the refurbishment, computer processing often filled in details where they did not exist, such as the ladder-like structures in the "jets" emerging from the core of the star, Eta Carinae (bottom picture). The Hubble repair immediately showed which of these so called "deconvolution artefacts" were spurious. With its two luminous clouds of gas, Eta Carinae is surely one of the most extraordinary objects yet seen by astronomers. Although now invisible to the naked eye, it was in the 1840s the second brightest star in the sky.

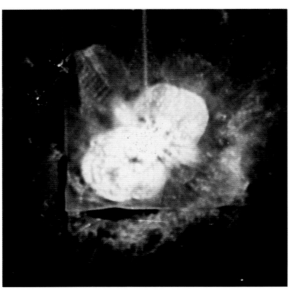

Optics Space Telescope Axial Replacement), would correct the telescope's myopia by bringing the light to a focus via a series of tiny mirrors each the size of a coin. The focused light would then enter the two spectrometers and the Faint Object Camera, which completed the telescope's scientific payload.

Ground controllers had scheduled five ambitious spacewalks, each six hours long, with two additional walks if needed. But even the most optimistic NASA managers wondered if all the objectives could be completed in one flight. Remarkably, with just a handful of minor problems, all the tasks were indeed achieved. Not only was Hubble's sight restored, so too was the credibility of NASA, which came as a pleasant antidote to a year otherwise fraught with bad news. The year 1993 had seen the search for extraterrestrial intelligence project termi-

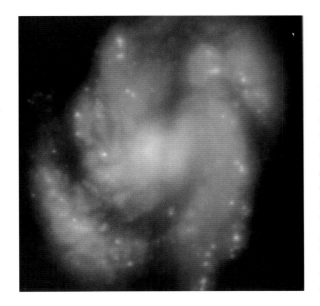

◄► **Hubble pay-off 2**
The obvious benefits of the Hubble "fix" are seen in the "before" and "after" shots of the same spiral galaxy, M100. The clarity of the dust lanes within the spiral arms is especially striking. The M100 galaxy is one of the brightest objects in the Virgo cluster of galaxies, and is similar in type to our own Milky Way galaxy.

nated, the space station — NASA's most vaunted project — heavily pruned, and the Mars Observer and a Landsat remote-sensing satellite both lost. A relieved President Clinton, in a phone call to the crew of *Endeavour*, agreed: "It shows the world we can still do what we want to do in space".

Global fix

A week after the Hubble Repair Mission, it was clear that the optical fix had been successful. On 18 December, astronomers gathered anxiously at the Space Telescope Science Institute (STScI) in Baltimore to await the first data from Wifpic 2, and breathed easy as calibration shots revealed straight-away that the corrective measures had worked. After further internal alignments of mirrors within the COSTAR instrument, the Faint Object Camera provided by the European Space Agency was ready for operations on New Year's Day. In the two weeks that followed, astronomers at the STScI worked around the clock to obtain further images which showed beyond all doubt that the telescope's blurred sight was now fully corrected. These, and other startling images, were revealed at a press conference held on 13 January 1994 at NASA's Goddard Space Flight Center, from where the tele-scope is controlled. Senator Barbara Mikulski, hold-ing a magnificent image of a galaxy that showed beyond all doubt the spherical aberration had been eradicated, declared unequivocally that "the trouble with Hubble is over".

"Faster, better, cheaper"

A few days after *Endeavour* returned to Earth, another significant event heralded a new direction in the future of planetary exploration. At the Applied Physics Laboratory of Johns Hopkins University, a few blocks from the STScI, scientists presented details of the first of the Discovery mis-sions. With a tendency for space missions to get ever more complex and unwieldy, NASA is now concentrating its efforts on "faster, better, cheaper" missions. The first concrete example of this new approach is the creation of small, relatively simple missions to be built, tested and operated indepen-dently of NASA. One of the first Discovery space-craft, it was announced, would be the Near Earth Asteroid Rendezvous (NEAR) mission, the first spacecraft designed to orbit an asteroid.

Studies of these minor planets have come a long way since the first of them, Ceres, was discovered on New Years' Day, 1801. Many thousands have now been found, and there are thought to be up to a million objects with a diameter of a kilometre or more orbiting in the region between Mars and Jupiter. Two of these bodies have been scanned from close range by the Galileo spacecraft on its extended journey through the asteroid belt. In October 1991, Galileo snapped asteroid Gaspra, and in September 1993, asteroid Ida. Both asteroids appeared more or less as expected: they are small and irregular bodies, lightly peppered with craters and relatively dark with little colour variation on their surfaces. In fact, they are remarkably similar in appearance to the Martian moons, Phobos and Deimos, giving weight to the idea that Mars cap-tured its satellites. The one real surprise came in the spring of 1994, when JPL announced that Ida itself has a kilometre-sized satellite, which has been given the name Dactyl. There had been indirect evi-dence that some asteroids were not lone objects, but here was the first proof.

These fleeting visits to asteroids by space probes are not the only means of investigating them. Collisions between asteroids are believed to be fre-quent, with larger bodies gradually being broken

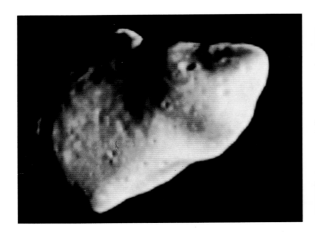

▲ Galileo views Gaspra
In October 1991, the Galileo spacecraft made a close pass of the asteroid Gaspra. The image colour was adjusted to appear, more or less, as it would to the naked eye. Gaspra has a rocky surface less grey than that of Earth's Moon, and measures 19 by 12 by 11 kilometres.

▲ Next step The first in a series of modestly priced Discovery spacecraft will hopefully be launched in late 1996. Seen here is an artist's impression of the near-Earth asteroid Eros, the target for the NEAR (Near Earth Asteroid Rendezvous) mission. Construction of the craft is already under way at the Johns Hopkins University in Baltimore.

◄ Gaspra close up
Colour enhancement of Galileo images obtained through violet, green and infrared filters revealed subtle details in the surface composition of asteroid Gaspra. Bluish areas represent regions of slightly higher reflectivity, perhaps indicating the presence of the mineral olivine, and seem to be associated with crisper craters and ridges. Slightly reddish areas seem to be coincide with lower-lying features, and are perhaps covered with the powdered rock material known as regolith.

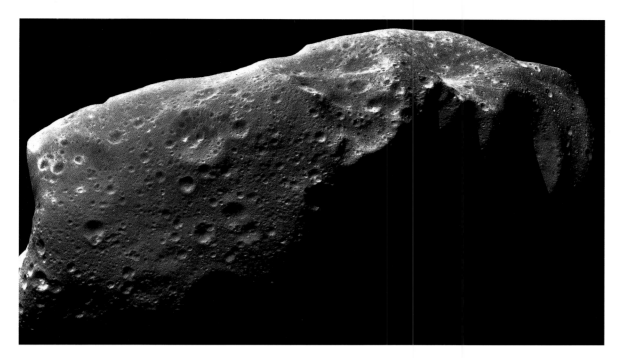

◄ **Asteroid Ida** In August 1993, Galileo made a close pass of asteroid Ida, which is 55 km in length and three times the size of Gaspra. Galileo's clear filter was used to produce this extremely sharp composite image, showing features smaller than 40 metres. The many smooth-sided craters dispelled earlier theories that Ida might be geologically youthful.

into smaller ones. Occasionally, a piece of debris from the asteroid belt lands on the Earth as a meteorite, surviving the fiery heat of its plunge through our atmosphere because of its relatively large size. For a handful of meteorites, whose falls were observed accurately, orbits have been calculated which show they originated in the asteroid belt. Most meteorites are rocky, which suggests that they were formed in the outermost layers of a planetary body. The largest known asteroid, Ceres, is more than 900 kilometres in diameter, large enough to have undergone differentiation, with the heavier elements, such as iron and nickel, sinking under the influence of gravity while the asteroid was still at least partly molten to form a core (see page 29). About five per cent of meteorites that have been actually seen to fall to Earth contain a mixture of iron and nickel, suggesting they were formed in the cores of such larger asteroids.

Another class of meteorites are darker, their colour resulting from the presence of carbon. Inside some are spherical grains of lighter rock known as chondrules, giving these meteorites the name carbonaceous chondrites. These are remarkable as they do not seem to have ever melted, which means they cannot have come from proto-planets that differentiated. They may represent the most ancient material in the asteroid belt; the only question is whether they come from pristine planetesimals, or from the surfaces of asteroids that did not melt. Clearly, we need more detailed observations, for the full story of the asteroid belt cannot be determined from a handful of observation made so far, which including spectroscopic and infrared studies. But a new mission, the NEAR probe, may tip the balance.

The NEAR spacecraft will execute a circuitous trajectory after its planned launch by a Delta 2 booster from Earth in February 1996. First, it will head out towards the main asteroid belt and pass close to the small asteroid Iliya, before arcing back towards the Earth and reaching the near-Earth asteroid Eros at the end of 1998. The near-Earth asteroids range in size from metres to kilometres, and some of them pass perilously close to us. Eros is one of the largest and its orbit is known very accurately, making it a good candidate for exploration. Thanks to a quirk of celestial mechanics, less energy is needed to reach Eros than to land on the surface of the Moon. NEAR's task will be to determine the mass, shape and composition of Eros, and to see if it has a magnetic field. The planned duration of the NEAR mission is one year.

The NEAR mission was only formally approved at the start of 1993, and this very rapid turnaround, from inception to launch in 1996, is one of the key precepts of the Discovery programme. By comparison, traditional space missions have taken anywhere from five to ten years in the defining stage before the hardware is actually built. The cost of each Discovery mission will be rigidly kept within $150 million — one-tenth of the cost of, for example, the Cassini mission to Saturn.

The full roster of Discovery missions under consideration includes spacecraft to our Moon and to all the inner planets; as well as missions with self-explanatory names like: Cometary Coma Chemical Composition probe, Comet Nucleus Tour, Earth Orbital Jovian Observer and a Solar Wind Sample Return. Time will only tell if the $150 million cost "ceiling" can actually be met, and likewise the tight

▼ NASA's cheap option
The Russian Proton rocket may be a means of meeting the $150 million budget for future planetary probes.

◀▼ Togetherness
NASA's main long-term goal is to build a space station in co-operation with Russia. In early February 1995, despite problems with a leaky fuel valve, the space shuttle Discovery successfully approached to within 12 metres of the Mir station, as a practice run for a full docking mission (artist's impression at left). In 1997, construction on Station Alpha is due to begin in earnest (below), although the usual financial uncertainties continue to plague the project, and cost overruns will impact on future planetary missions.

deadlines for furnishing hardware. One criticism is that such low cost budgets do not permit missions much beyond Jupiter, but this may not necessarily be the case. JPL is once more in the vanguard, for its engineers plan to reach the only world they have yet to see from close range, Pluto.

The cannonball run

There can be few space missions that have been directly inspired by a postage stamp, but when the tenth in a series of stamps issued by the US postal service in 1991 depicted Pluto with the legend "NOT YET EXPLORED", it was a direct challenge to planners in JPL's Mission Design section. In particular, two young scientists, Rob Staehle and Stacey Weinstein, were moved to act and started to think about innovative ways of reaching the outermost world. Out of their deliberations came an idea that chimed with NASA's new "faster, better, cheaper" approach to space missions.

The proposed Pluto Fast Fly-by spacecraft has been referred to as a cannonball with a camera and radio transmitter. Planned to weigh less than 162 kilograms when fully fuelled and to stand only 1.2 metres high, it will look like a miniaturized version of the Voyager spacecraft. But looks are deceptive, for it represents a new departure in the construction of planetary spacecraft. The craft will operate on only 60 watts of power – the same as a domestic light bulb – drawn from a radioisotope thermoelectric generator. On-board systems will be adapted from existing hardware to keep costs and complexity down, though will be updated where possible to include the latest in microtechnology.

The Pluto-bound spacecraft will orientate itself in space with star trackers, originally developed by the military for missile guidance. Although these instruments need to work for only 15 minutes on Earth, they will have to operate for over a decade in deep space. Scientific instruments, co-ordinated by a central computer similar to a domestic PC, will include a camera, as well as infrared and ultraviolet spectrometers for mapping the surfaces of Pluto and Charon. Radio transmissions will also be used to probe Pluto's atmosphere.

Speed is of the essence, for time is against the mission planners; at some point before 2020, Pluto will move farther out from the Sun along its oblique

orbit and atmospheric activity will cease as the atmospheric gases freeze solid until the twenty-third century. Of equally pressing concern is the fact that Pluto and Charon will begin to cast shadows upon each other from 2005 onwards, because of the geometry of their orbits. As time goes on, the duration of the shadows will increase, making the mapping of both worlds difficult.

If the mission is launched in 1999, aboard a Titan IV/Centaur rocket, it would take just seven years to reach Pluto. If another launcher is used — a Russian Proton has been suggested — it could take longer. Although a gravity assist with Jupiter could be used, the launch would have to be delayed until 2001 for Jupiter to be in the right part of the sky; the total flight time would then be over a decade. But only by taking a fast track, will Pluto and Charon be seen in their full glory.

If it goes ahead, the Pluto Fast Fly-by will usher in a new era of lower cost missions which may eventually be employed to explore all the planets. It

▲▼ Past and future
A replica of the explorer Magellan's ship sails past the spacecraft Magellan's launchpad (above). A backlog of probes has built up since the Challenger accident, and planetary exploration continues to be hampered by the high cost of the space shuttle as NASA's primary launch system.

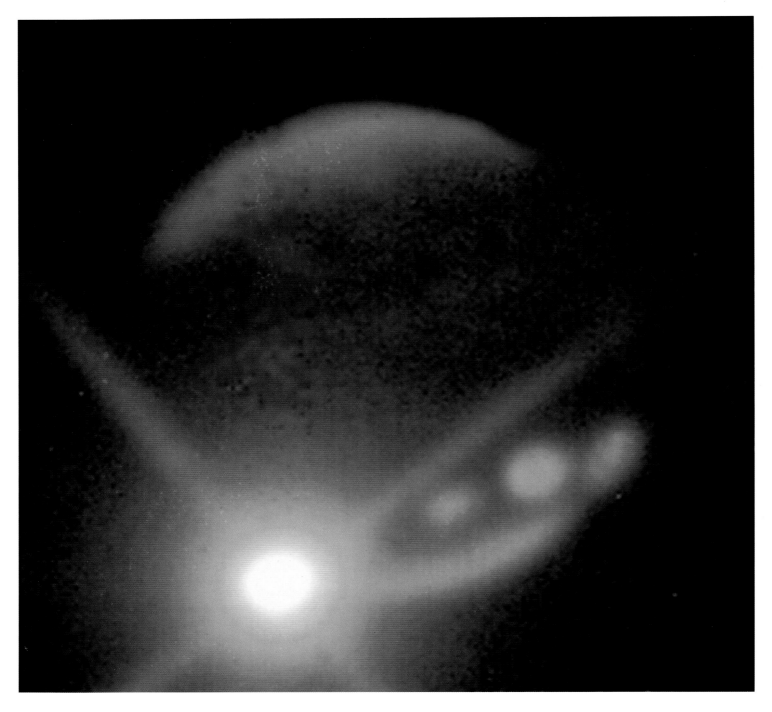

▲ **Flashpoint** *More than 20 fragments of Comet Shoemaker-Levy 9 collided with Jupiter between 16 and 22 July 1994. This infrared image, from the 2.3 metre Mt. Stromlo Observatory in Australia, shows the fireball created 13 minutes after Impact K at 10.18 GMT on 18 July. Earlier impact sites are visible to the right of the 20,000 km-wide fireball.*

is quite obvious that the direction of planetary exploration has now changed, certainly as far as the United States is concerned. The standard argument for "big" missions is that they can make a wide range of observations with only one overhead cost for the launch. But the view now is that the more complex the mission, the more susceptible it is to failure, and a very costly failure at that.

It is not yet clear whether large-scale space endeavour has a future. One of the axioms of big scientific projects is that international co-operation is the best way forward. Yet, the attraction of sharing the burden of cost between nations has to be weighed against the additional administrative and engineering complications. A prime example is the international space station, which has been in planning for over ten years and has progressed only in fits and starts, repeatedly suffering near-cancellation. Its latest incarnation is Station Alpha, which will involve Russian hardware that has already been tested during the long-term *Mir* missions. But the role of European, Canadian and Japanese partners is not clear, and the escalating cost of the project has eaten into other budgets. Perhaps it is too much to hope that the apparently esoteric enterprise of planetary exploration will be able to hold its place in the future against the more politically attractive goal of putting people in space.

◄ **Scars of battle** *Dark brown spots mark the trail left by Comet Shoemaker-Levy 9, after it smashed into Jupiter in July 1994. At centre, near the small white spot on Jupiter, is the star-shaped site of Impact H (19.26 GMT on 18 July); at lower right are the three sites of Impact Q₁ (19.59 GMT on 20 July), Impact R (05.22 GMT, 21 July) and the merged Impact D/G (17/18 July).*

The best is yet to come

So the future of planetary exploration will remain, as it has been for over 20 years, uncertain. There is still much to be learned about the Solar System. Only one hemisphere of Mercury has been mapped so far, and little fine scale detail has been seen on the surface of Venus. The important polar regions of the Moon have only just been observed, and it will be quite a while before a consistent geochemistry of the whole lunar surface is accomplished. The same could be said of Mars, where fundamental questions about the planet's evolution and the role of water remain unresolved. Although fantastically successful, the Voyager missions have returned only tantalizing snapshots of the outer gas giant planets; more detailed investigations are required to piece together a better understanding of their weather and climate, and of the geological evolution of their many varied moons. We have our first close glimpses of comets and asteroids, and they seem to be related to each other and to the recently discovered Kuiper Belt objects out beyond the orbit of Neptune, but more detailed examination is needed. The realization that these small, icy worlds may hold important clues to the Solar System's birth warrants urgent investigation.

We can be consoled, however, by the fact that whatever happens, the best is yet to come.

INDEX

Italic page numbers refer to picture captions.

PICTURE CREDITS

Abbreviations: L Left, R Right, T Top, B Bottom, C Centre; BMDO Ballistic Missile Defense Organization, LLNL Lawrence Livermore National Laboratory, NRL Naval Research Laboratory, SPL Science Photo Library, STScI Space Telescope Science Institute, USGS United States Geological Service.

Bradford Smith/University of Arizona 128 T. Cornell University 62 BL. ESA 12 B, 26, 37, 41, 48, 49 B, 51 B, 124, 125 TL, 152; /NASA 24 B, 122 R. Hughes Aircraft Co. 74 TR. Johns Hopkins University 166 TR. Lockheed Space and Missile Company/ISAS 31 T, 62 T. Mark Robinson, University of Hawaii 9 T, 96 B, 96 T. NASA 160; /Ames 73, 74 B, 74 TL, 75 T, 81 TR, 100 L, 100 R, 104 TR, 105 BL; /C.R. O'Dell/Rice University 28 TL, 28 TR; /Center for North Atlantic Weather Forecasting 46 BL; /JPL 7, 10 T, 13, 14 B, 14 T, 15 B, 15 T, 18 B, 19 C, 20 C, 20 TL, 20 TR, 21 B, 23 R, 23 TL, 25 B, 30 L, 50 B, 51 TR, 56 B, 60, 62 BR, 63 B, 63 T, 65 BL, 65 C, 65 T, 68, 70, 71, 72 B, 72 T, 75 B, 76 B, 76 T, 77 T, 78 B, 78 T, 79 B, 79 T, 80 T, 81 BL, 81 BR, 81 TL, 84 BR, 86 B, 86 T, 87 B, 87 T, 88 T, 94 B, 95 T, 98, 101 B, 102 B, 102 T, 105 T, 106 B, 106 BC, 106 T, 106 TC, 107 L, 107 R, 108 C, 109 B, 110 B, 110 T, 111, 112, 114 B, 114 TL, 114 TR, 115, 116 B, 116 T, 117 B, 117 T, 118 BL, 118 TL, 118 TR, 119 B, 120 B, 120 TL, 120 TR, 121 B, 121 T, 125 B, 125 CR, 125 TR, 126, 128 B, 129 BL, 129 BR, 129 T, 130 B, 130 T, 131 B, 131 T, 132 L, 132 R, 133 B, 133 C, 133 T, 135, 136, 138, 139 B, 139 T, 140 R, 141 B, 141 T, 142 B, 142 T, 143 L, 143 R, 144 B, 144 C, 144 T, 145, 146 B, 146 T, 158 T, 166 B, 166 TL, 167; /JPL/Caltech 66 B; /JPL/CNES 49 T; /JPL/Joe Waters 46 T; /JSC 6, 8 B, 10 B, 14 R, 18 C, 18 T, 32 T, 33, 34, 36, 40, 42 B, 42 TR, 44, 45 T, 46 C, 47, 50 T, 51 TL, 52, 54, 55 B, 55 T, 58, 134 B, 134 C, 134 T, 162, 163 B, 163 T, 168 B, 168 TR, 169 B, 169 T; /JSC/Pat Rawlings 95 B; /STScI 21 T, 28 CL, 28 TC, 85 BL, 103 T, 105 BR, 119 T, 156 B, 156 T, 158 B, 164, 164 T, 165 TL, 165 TR; /Washington University 22, 84 TL, 90 B, 90 T, 91. National Remote Sensing Centre 46 BR. Nicholas Booth 12 T, 101 T, 118 BR, 122 CL, 122 TL. NPO Energia 168 TL. NRL/BMDO/LLNL 11, 53, 57, 59. Pat Rawlings/SAIC 149, 159. Planetary Visions 45 C. Space Research Institute (IKI), Moscow 18 B. SPL/A.E. Potter & T.H. Morgan/NASA 66 T; /Chris Butler 67; /David P. Anderson, SMU/NASA 8 T; /David Hardy 29, 30 R, 103 B; /David Hathaway/NASA 32 B, 16; /Dr Gene Feldman, NASA GSFC 38; /Dr. Michael J. Ledlow 65 BL; /ESA 154, 155; /J. Tennyson & S. Miller/University College London 104 TL; /Max-Planck-Institut für Aeronomie/David Parker 150; /MSSSO, ANU 170; /NASA 23 BL, 56 T, 64, 75 C, 77 B, 80 B, 84 BL, 92 T, 148; /NASA/Space Telescope Science Institute 171; /NOAO 31 B; /Novosti Press Agency 94 T; /Roger Ressmeyer, Starlight 140 L; /USGS 85 BR, 88 B, 89 B, 89 TL, 92 B, 109 T; /USGS/NASA 24 T, 108 T. USGS 25 T, 83, 84 TR, 93; /David H. Harlow 42 TL. Dr. Warren Moos, Johns Hopkins University 24 C.

DATE DUE